Ordinary Differential Equations

SIAM's Classics in Applied Mathematics series consists of books that were previously allowed to go out of print. These books are republished by SIAM as a professional service because they continue to be important resources for mathematical scientists.

Editor-in-Chief
Gene H. Golub, *Stanford University*

Editorial Board
Richard A. Brualdi, *University of Wisconsin-Madison*
Herbert B. Keller, *California Institute of Technology*
Ingram Olkin, *Stanford University*
Robert E. O'Malley, Jr., *University of Washington*

Classics in Applied Mathematics

C. C. Lin and L. A. Segel, *Mathematics Applied to Deterministic Problems in the Natural Sciences*

Johan G. F. Belinfante and Bernard Kolman, *A Survey of Lie Groups and Lie Algebras with Applications and Computational Methods*

James M. Ortega, *Numerical Analysis: A Second Course*

Anthony V. Fiacco and Garth P. McCormick, *Nonlinear Programming: Sequential Unconstrained Minimization Techniques*

F. H. Clarke, *Optimization and Nonsmooth Analysis*

George F. Carrier and Carl E. Pearson, *Ordinary Differential Equations*

Leo Breiman, *Probability*

R. Bellman and G. M. Wing, *An Introduction to Invariant Imbedding*

Abraham Berman and Robert J. Plemmons, *Nonnegative Matrices in the Mathematical Sciences*

Olvi L. Mangasarian, *Nonlinear Programming*

*Carl Friedrich Gauss, *Theory of the Combination of Observations Least Subject to Errors: Part One, Part Two, Supplement.* Translated by G. W. Stewart

Richard Bellman, *Introduction to Matrix Analysis*

U. M. Ascher, R.M.M. Mattheij, and R.D. Russell, *Numerical Solution of Boundary Value Problems for Ordinary Differential Equations*

K. E. Brenan, S. L. Campbell, and L. R. Petzold, *Numerical Solution of Initial-Value Problems in Differential-Algebraic Equations*

Charles L. Lawson and Richard J. Hanson, *Solving Least Squares Problems*

*First time in print.

Ordinary Differential Equations

George F. Carrier
Harvard University

Carl E. Pearson
University of Washington

Society for Industrial and Applied Mathematics
Philadelphia

Copyright © 1991 by the Society for Industrial and Applied Mathematics.

This SIAM edition is an unabridged, corrected republication of the work first published by Blaisdell Publishing Company, Waltham, MA, 1968.

10 9 8 7 6 5 4 3 2

All rights reserved. Printed in the United States of America. No part of this book may be reproduced, stored, or transmitted in any manner without the written permission of the publisher. For information, write the Society for Industrial and Applied Mathematics, 3600 University City Science Center, Philadelphia, PA 19104-2688.

Library of Congress Cataloging-in-Publication Data

Carrier, George F.
    Ordinary differential equations / George F. Carrier, Carl E. Pearson.
       p. cm. -- (Classics in applied mathematics :6)
    Originally published: Waltham, Mass. : Blaisdell Pub. Co., 1968.
    Includes bibliographical references.
    ISBN 0-89871-265-3
    1. Differential equations.   I. Pearson, Carl E.   II. Title.
III. Series.
QA372.C33 1991
515'.35--dc20   91-2642

siam. is a registered trademark.

PREFACE

The material in this book is not a conventional treatment of ordinary differential equations. It does not contain the collection of proofs commonly displayed as the foundations of the subject, nor does it contain the collection of recipes commonly aimed at the scientist or engineer. Rather, in a way which requires little or no previous acquaintance with the subject, it contains a sequence of heuristic arguments, illustrative examples and exercises, which serve to lead the reader towards the invention, generalization, and usage of the techniques by which solutions of differential equations can be constructed. Above all, we hope, the reader can gain a perspective concerning the extent to which methods which lead *in principle* to the solution of a given problem actually lead to a *useful* description of that solution.

Our purpose is to offer an alternative to the almost "rote" approach, in which the standard categories of differential equations, accompanied by routine problem sets, are systematically listed. We firmly believe that the present approach is one that should be encountered, at least once, by mathematicians, users of mathematics, and those who are merely curious about mathematics; we hope that members of all three sets will find the presentation stimulating.

We consider the exercises to be an essential part of the text. They extend, amplify, and provide perspective for the text material. (In rare cases, the statement of a problem is left deliberately incomplete, so as to give the reader some scope for ingenuity.) If this book is used to accompany a course of lectures, one technique would be to assign text reading and exercises beforehand, to be followed by classroom discussion and amplification. Alternatively, special problem sessions could be included.

In reprinting this book, we have taken the opportunity to correct an (embarrassingly large!) number of misprints, and to clarify certain aspects of the presentation. As in the original printing, the attitudes and approaches in this book remain solely the responsibility of the authors; however, we gratefully acknowledge the efforts of the many colleagues who have struggled to reform our viewpoints. In this connection, we are particularly appreciative of the suggestions provided by Bob Margulies and Frank Fendell, each of whom was kind enough to read portions of the present manuscript.

G.F.C.
C.E.P.

CONTENTS

INTRODUCTION   1

Chapter 1   First-order linear differential equations   4

1.1   First-order linear ordinary differential equations   4
1.2   Problems   5
1.3   An illustrative sublimation problem   7
1.4   Problems   8
1.5   The nonhomogeneous equation   9
1.6   Problems   11
1.7   A nonlinear equation   13
1.8   Problems   13

Chapter 2   First-order linear difference equations   16

2.1   Problems   18

Chapter 3   Second-order differential equations   20

3.1   Problems   25
3.2   The homogeneous problem   28
3.3   Problems   29
3.4   Operators which can be factored   30
3.5   Problems   31

Chapter 4   Power-series descriptions   33

4.1   Problems   35
4.2   A more recalcitrant illustrative problem   36
4.3   Problems   38
4.4   Singular points   39
4.5   Problems   40
4.6   Singular points—continued   41
4.7   Problems   42

Chapter 5   The Wronskian   43

5.1   Problems   45

Chapter 6   Eigenvalue Problems   47

6.1   Problems   50
6.2   Fourier series   52
6.3   Problems   54
6.4   A special example   55
6.5   Problems   57

Chapter 7 The second-order linear nonhomogeneous equation 59

7.1 Problems 61
7.2 Green's functions 64
7.3 Problems 68
7.4 Nonhomogeneous boundary conditions 70

Chapter 8 Expansions in eigenfunctions 72

8.1 Problems 73
8.2 Two approximation methods 75
8.3 Problems 77

Chapter 9 The perturbation expansion 79

9.1 Problems 82
9.2 An eigenvalue problem 82
9.3 Problems 85

Chapter 10 Asymptotic series 86

10.1 Problems 88
10.2 An elementary technique 89
10.3 Problems 90
10.4 Another technique 91
10.5 Problems 93
10.6 Asymptotic expansions in a parameter 93
10.7 Problems 95

Chapter 11 Special functions 96

11.1 The error function 96
11.2 The gamma function 98
11.3 Bessel functions of integral order 100
11.4 Bessel functions of nonintegral order 103
11.5 The Airy functions 104
11.6 The Legendre polynomials 105

Chapter 12 The Laplace transform 107

12.1 Problems 107
12.2 Transform of a derivative 110
12.3 Problems 111
12.4 The convolution integral 114
12.5 Problems 115
12.6 Asymptotic behavior 116
12.7 Problems 117
12.8 Partial differential equations 118
12.9 Problems 119

Chapter 13 Rudiments of the variational calculus   121

13.1 Problems   124
13.2 The solution of differential equations   125
13.3 Problems   128

Chapter 14 Separation of variables and product series solutions of partial differential equations   130

14.1 The heat equation   130
14.2 Problems   133
14.3 Nonhomogeneous boundary conditions   134
14.4 Problems   135
14.5 Laplace's equation   136
14.6 Problems   138
14.7 Another geometry   139
14.8 Problems   141
14.9 The Helmholtz equation   142

Chapter 15 Nonlinear differential equations   145

15.1 Problems   147
15.2 Second-order equations; the phase plane   148
15.3 Problems   151
15.4 Singular points   152
15.5 Problems   155

Chapter 16 More on difference equations   157

16.1 Second-order equations   157
16.2 Problems   159
16.3 The general linear equation   160
16.4 Problems   162
16.5 Generating functions   163
16.6 Problems   165
16.7 Nonlinear difference equations   166
16.8 Problems   168
16.9 Further properties of difference equations   170

Chpater 17 Numerical methods   174

17.1 Problems   178
17.2 Discretization error; some one-step methods   179
17.3 Problems   182
17.4 Runge-Kutta method   183
17.5 Problems   185
17.6 Multistep processes; instability   185
17.7 Problems   188
17.8 Two-point boundary conditions; eigenvalue problem   189
17.9 Problems   191

Chapter 18 Singular perturbation methods   193

18.1 The boundary layer idea   193
18.2 Problems   199
18.3 Interior layers and a spurious construction   202
18.4 Problems   205
18.5 The turning-point problem   205
18.6 Problems   209
18.7 Two-timing   210
18.8 Problems   216

Index   219

Introduction

Differential equations are merely equations in which derivatives of an unknown function appear. *Ordinary* differential equations are those in which there appear derivatives with respect to only one independent variable. A simple example of an ordinary differential equation is

$$\frac{df(x)}{dx} = \sigma f(x), \tag{I.1}$$

where σ is a constant. By a *solution* of this differential equation on an interval $a < x < b$ is meant a function $f(x)$ for which $f(x)$ and $df(x)/dx$ are defined* at each point x in the interval and for which the equation reduces to an identity for each value of x in that interval. Such a function is said to *satisfy* the differential equation.

Using interchangeably the notation $df(x)/dx \equiv df/dx \equiv f'(x) \equiv f'$, other examples of ordinary differential equations are

$$[xg'(x)]' + \alpha x g(x) = 0, \tag{I.2}$$

$$u''' + uu' = 0, \tag{I.3}$$

$$w(d^2w/dx^2) + \beta \sin w = e^x, \tag{I.4}$$

where α and β are constants.

There is an enormous variety of questions that can be asked in the study of a differential equation. Thus, one might ask, with reference to Equation (I.1):

(a) What function or functions (if any) satisfy Equation (I.1) when $\sigma = 1$?

---

* On occasion, we will relax this requirement somewhat and permit a solution to be undefined at one or more critical points. Such a less restricted solution could be termed *weak*, in contrast with the *strong* solution defined above.

(b) For what values of the number σ does a function $f(x)$ exist which satisfies Equation (I.1)?
(c) How many functions satisfy Equation (I.1) for each value of σ?
(d) How many functions $f(x)$ are there for which $f(x)$ and $f'(x)$ are continuous at each real x, for which $f(0) = 0$, and for which Equation (I.1) is satisfied? What are they?
(e) How many functions $f(x)$ are there for which $f(x)$ and $f'(x)$ are continuous at each real x, for which $f(1) = 1$, and for which Equation (I.1) is satisfied? What are they?
(f) How many functions $f(x)$ are there for which $f(x)$ and $f'(x)$ are continuous at each real x, for which $f(1) = 1$, for which $f(2) = 2$, and for which Equation (I.1) is satisfied? What are they?
(g) How many functions $f(x)$ are there for which $f(x)$ and $f'(x)$ are continuous at each real x except $x = 0$, for which $f(1) = 1$, for which $f(-1) = -1$, for which $f(0) = 0$, and for which Equation (I.1) is satisfied at each x except $x = 0$? What are they?

It is a familiar fact to those who have studied the calculus that any constant multiple of $e^{\sigma x}$ satisfies Equation (I.1) at each point x. It is readily shown that no other function $f(x)$ will do so. To see this, we note that $e^{\sigma x}$ is never zero, so that we can define $u(x)$ for all x by

$$f(x) = e^{\sigma x} u(x). \tag{I.5}$$

Substituting this into Equation (I.1), we obtain

$$u'(x) = 0, \tag{I.6}$$

from which

$$u(x) = \text{const.} \tag{I.7}$$

Thus, from Equation (I.5), the only possible solution of Equation (I.1) is a constant multiple of $e^{\sigma x}$. Armed with this information, the student should verify that the answers to the foregoing questions are:

(a) All functions of the form $f(x) = Ae^x$, where A is any constant.
(b) For every value of σ.
(c) Infinitely many.
(d) One function. $f(x) = 0$.
(e) One function. $f(x) = e^{(x-1)\sigma}$.
(f) None, unless $\sigma = \ln 2$; then one. $f(x) = 2^{x-1}$.
(g) One function.

$$f(x) = \begin{cases} e^{\sigma(x-1)}, & x > 0, \\ 0, & x = 0, \\ -e^{\sigma(x+1)}, & x < 0. \end{cases}$$

A broader set of questions is:

(a) How many constraints must accompany a differential equation in order to imply a unique solution?
(b) What role is played by the prescription of the domain (in x) on which the equation is to be satisfied?
(c) How do differential equations arise outside the classroom?
(d) How may differential equations be classified conveniently?
(e) For which classes of differential equations and additional constraints can useful theorems regarding the existence and uniqueness of solutions be found?
(f) What techniques for finding solutions of differential equations are recurrently useful?
(g) What differential equations are encountered so often that each merits intensive study?

We shall address ourselves to these and many other questions, giving a strong preference to those whose answers are vital to the "user of mathematics."

First-Order Linear Differential Equations | 1

Perhaps the most important classification to be identified is that which distinguishes linear differential equations from nonlinear ones. A linear ordinary differential equation has the form

$$\sum_{n=0}^{N} a_n(x) \frac{d^n f}{dx^n} = G(x), \tag{1.1}$$

where $G(x)$ and the $a_n(x)$ are prescribed functions. Those which cannot be cast in this form are nonlinear. Equations (I.1) and (I.2) are linear; (I.3) and (I.4) are nonlinear. Equation (1.1) is said to be of Nth order. It is nonhomogeneous unless $G(x)$ is zero for each value of x under consideration; when $G(x)$ *is* zero for all such x, Equation (1.1) is homogeneous.

Most of this text will be concerned with linear differential equations because they are of great importance, because much of the study of nonlinear equations requires an understanding of linear equations, and especially because our understanding of linear equations is in far better shape than our understanding of nonlinear ones.

1.1 First-Order Linear Ordinary Differential Equations

In accord with Equation (1.1), each first-order linear homogeneous equation has the form

$$a_1(x)f'(x) + a_0(x)f(x) = 0. \tag{1.2}$$

Any function $f(x)$ which satisfies Equation (1.2) can be found as follows. We divide each term in Equation (1.2) by $a_1(x)f(x)$ to obtain

$$\frac{f'(x)}{f(x)} = -\frac{a_0(x)}{a_1(x)}, \qquad (1.3)$$

noting that Equation (1.3) is meaningless at points where $a_1(x) = 0$ and/or $f(x) = 0$. It is also meaningless at any point x where a_0 or a_1 is not defined. At all other points x, Equation (1.3) implies that

$$\ln f(x) = -h(x)$$

and

$$f(x) = e^{-h(x)}, \qquad (1.4)$$

where $h(x)$ is any function whose derivative is $a_0(x)/a_1(x)$. Throughout this book we will use the notation $\int^x g(t)\,dt$ to denote "any function whose derivative is $g(x)$."

A few examples will illustrate a variety of directions of inquiry to which these results may lead. THE READER WHO FAILS TO DO THESE EXERCISES AND THEIR COUNTERPARTS THROUGHOUT THE TEXT WILL MISS 78% OF THE VALUE OF THE BOOK!

1.2 Problem

Find all of the solutions of each of the following equations. On what interval is each of the solutions defined?

(a)
$$\phi'(x) + (\sin x)\phi(x) = 0; \qquad (1.5)$$

(b)
$$\phi'(x) + p(x)\phi(x) = 0, \qquad (1.6)$$

where

$$p(x) = \begin{cases} 0, & x \leq 0, \\ \sin x, & x > 0; \end{cases}$$

(c)
$$\phi'(x) + q(x)\phi(x) = 0, \qquad (1.7)$$

where

$$q(x) = \begin{cases} 0, & x \le 0, \\ \cos x, & x > 0; \end{cases}$$

(d)
$$\phi'(x) + r(x)\phi(x) = 0, \tag{1.8}$$

where
$$r(x) = \begin{cases} 1, & x \text{ irrational}, \\ 0, & x \text{ rational}; \end{cases}$$

(e)
$$\phi'(x) + s(x)\phi(x) = 0, \tag{1.9}$$

where
$$s(x) = \begin{cases} x^{-1}, & |x| > 0, \\ 0, & x = 0; \end{cases}$$

(f)
$$(x^2 - 1)\phi'(x) + (x - 1)\phi(x) = 0; \tag{1.10}$$

(g)
$$(1 - x^2)\phi' + \phi = 0; \tag{1.11}$$

(h)
$$(a - x)^2 \phi'(x) + \phi(x) = 0. \tag{1.12}$$

How are the answers to the foregoing questions modified when the definition of a solution does not include the requirement that $\phi'(x)$ be continuous at every x in the interval?

The reader who has thoroughly studied these exercises will now realize that:
(1) The continuity, or lack thereof, of the solutions of differential equations depends on the character of the coefficients in the equation.
(2) Ordinarily, in order to imply a uniquely defined solution to a differential equation, further constraints must be imposed.
(3) The technique used in the Introduction to show that constant multiples of $e^{\sigma x}$ are the only functions which satisfy Equation (I.1) can also be used for the same purpose in the study of most linear homogeneous first-order ordinary differential equations.

(4) Differential equations in which the coefficients are too pathological don't have nontrivial solutions (a "trivial solution" $f(x)$ is that for which $f(x) \equiv 0$).

1.3 An Illustrative Sublimation Problem

Differential equations arise in a variety of contexts of which the following is not atypical.

Accepting the observational fact that paradichlorobenzene (moth repellent), when in a room at 20° centigrade, sublimes from the solid to the gaseous state at the rate of k cubic centimeters of solid per second per square centimeter of exposed solid surface, we would like to predict what radius a mothball will have at any subsequent time if it was observed to be spherical with a $\frac{1}{2}$-cm radius at noon on February 5 and was thereafter in a room at $20°C$.

We denote the radius of the ball (measured in cm) at t sec past noon of February 5 by the function $r(t)$. Since the exposed surface area, A, at time t is

$$A = 4\pi r^2(t),$$

since the volume, V, of the ball is

$$V = 4\pi r^3(t)/3,$$

and since the rate of decrease of volume (i.e. the decrease in volume per unit time) at time t is

$$\text{decrease rate} = -\frac{dV(t)}{dt},$$

the foregoing observational fact implies that

$$-\frac{dV}{dt} = kA.$$

That is,

$$-4\pi r^2(t) r'(t) = k 4\pi r^2(t). \tag{1.13}$$

It follows that, for all values of t less than that for which $r(t) = 0$, we may divide each side of Equation (1.13) by $4\pi r^2(t)$ to obtain

$$r'(t) = -k. \tag{1.14}$$

Our other observational starting point (the mothball had radius $\frac{1}{2}$ cm at noon) implies that

$$r(0) = \tfrac{1}{2}. \tag{1.15}$$

It is a simple matter to conclude that the only function $r(t)$ which satisfies Equation (1.14) for $t > 0$ and obeys Equation (1.15) is

$$r(t) = \tfrac{1}{2} - kt. \tag{1.16}$$

It is also a simple matter to conclude that our mothball will have disappeared at $t = 1/2k$ (this result provides us with an experimental method for determining k) and that, in the context of this problem, Equation (1.16) is of interest only in the domain $0 \leq t \leq 1/2k$.

However! The most important thing to note is that the mathematical model of the original question consisted of a differential equation *and* a boundary condition (or an initial condition) and that these *two* mathematical constraints on $r(t)$ implied a unique description of $r(t)$. It is typical of the mathematical models of problems that arise from the sciences that they consist of a differential equation (or equations) and an appropriate number of boundary conditions; in particular, it is typical of those which give rise to a first-order differential equation that there should be one boundary condition. This is a very happy fact because whenever, over the range of interest of the independent variable, the coefficients are nowhere too pathological, one boundary condition is precisely what is needed to render the solution unique. In particular, the student may verify that the appending of one boundary condition at an appropriate value of x to any of Equations (1.6), (1.7), (1.9), (1.10), (1.11), and (1.12) will imply a unique choice of $\phi(x)$ in each case, again in an appropriate region. Furthermore, there is no better way for the student to *begin* to see how the "appropriate region" and "appropriate values of x" depend on the degree of pathology exhibited by the coefficients of the differential equation.

The following problems illustrate the manner in which some differential equations arise and provide for further skill acquisition by the reader.

1.4 Problems

1.4.1 What are the cartesian co-ordinates $(x, y(x))$ of the points on the curve whose slope at x, y is a fixed multiple, σ, of the product of x and y? The curve should pass through the point (α, β).

1.4.2 A typical clinical thermometer will increase its reading by $1°$ C when approximately 10^{-2} calories of heat are added to the bulb and its contents. When in use, heat is transferred to this portion of the instrument at a rate proportional to the discrepancy between its temperature and that of its environment. Find the differential equation governing the temperature reading as a function of time, and solve it. Estimate the heat transfer rate, using any experience you have (or experiments you are willing to perform) with a clinical thermometer.

1.4.3 As is schematically illustrated in Figure 1.1, the area, $A(L)$, under that portion of the curve, $y(x)$, which lies between the line $x = 0$ and the line $x = L$, is to be such that, for each L,

(a) $\quad A(L) = y(L),$
(b) $\quad A(L) = y(L) \sin L,$
(c) $\quad A(L) = (1 + L)y(L),$
(d) $\quad A(L) = y(L)/L.$

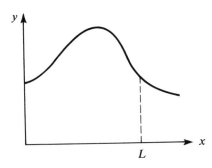

Figure 1.1 Curve $y(x)$ for Problem 1.4.3.

In each of (a) through (d), is there any curve which meets the requirement? Is it uniquely determined?

1.5 The Nonhomogeneous Equation

We now turn our attention to the first-order linear nonhomogeneous equation

$$a_1(x)u' + a_0(x)u = g(x). \tag{1.17}$$

We recall that the solution of the homogeneous counterpart of Equation (1.17) was given by

$$f(x) = e^{-h(x)},$$

where

$$h(x) = \int^x a_0(t)/a_1(t)\, dt.$$

We note, further, that

$$e^{h(x)}[a_1(x)u'(x) + a_0(x)u(x)] = a_1(x)[e^{h(x)}u(x)]'. \quad (1.18)$$

Thus, we may use the solution of the homogeneous equation to rewrite the nonhomogeneous equation in the form

$$[e^{h(x)}u(x)]' = g(x)e^{h(x)}/a_1(x), \quad (1.19)$$

and $u(x)$ can be found by integrating both sides of Equation (1.19) to obtain

$$u(x) = e^{-h(x)} \int^x [g(t)e^{h(t)}/a_1(t)] \, dt \quad (1.20)$$

which is valid in any x-interval that does not contain a zero of $a_1(x)$. This simple technique is the only one you will ever need to solve first-order linear ordinary differential equations.

The reader should note that any allowable choice of $h(x)$ differs from any other allowable choice by a constant (call it μ) and that this constant has no effect on the value of $u(x)$ as given by Equation (1.20). Thus, only the $\int^x$ process in Equation (1.20) introduces any essential indeterminacy; we observe that all functions satisfying Equation (1.20) differ from each other by a constant multiple of $e^{-h(x)}$. If we denote by $Q(x)$ a particular choice of

$$\int^x [g(t)e^{h(t)-h(x)}/a_1(t)]dt,$$

it follows that any solution of Equation (1.17) can be written as

$$u(x) = Ae^{-h(x)} + Q(x), \quad (1.21)$$

where A is a constant. A frequently used vocabulary identifies $Q(x)$ as a *particular solution* of Equation (1.17), and a conventional phrasing of Equation (1.21) is: *any solution of Equation (1.17) consists of the sum of a particular solution and a multiple of the complementary solution.* The latter, clearly, is merely another term for a *nontrivial solution of the homogeneous equation.**

As an illustrative example, we consider the equation

---

*At this point, the reader should readily anticipate and verify that if $P(x)$ is a particular solution of Equation (1.1) and if $u_j(x)$ (with $j = 1, 2, \cdots, R$) are solutions of the homogeneous equation, that is, Equation (1.1) with $G(x) = 0$, then with any numbers A_j, $P(x) + \sum_{j=1}^{R} A_j u_j(x)$ is also a solution of Equation (1.1). This statement is frequently referred to as the principle of superposition for linear equations. The reader should verify also that the statement does not apply to solutions of nonlinear equations.

$$\phi'(x) + x\phi(x) = xe^{x^2}, \qquad (1.22)$$

where we seek that function $\phi(x)$ for which

$$\phi(0) = 1.$$

The foregoing procedure gives, in successive steps,

$$[e^{x^2/2}\phi(x)]' = xe^{3x^2/2}, \qquad e^{x^2/2}\phi(x) = \text{any constant} + \tfrac{1}{3}e^{3x^2/2},$$

and *one* particular choice for $\phi(x)$ is

$$Q(x) = \tfrac{1}{3}e^{x^2}. \qquad (1.23)$$

Any solution, $\phi(x)$, is given by

$$\phi(x) = Q(x) + Ae^{-x^2/2} = \tfrac{1}{3}e^{x^2} + Ae^{-x^2/2}. \qquad (1.24)$$

In the foregoing problem, the formal process may not have been necessary; the perceptive student who knew that

$$\frac{d}{dx}(Be^{cx^2}) = 2xBce^{cx^2}$$

might well have recognized that the $Q(x)$ of Equation (1.23) was a solution of Equation (1.22); he could have chosen this as *the particular solution*, and he could have written Equations (1.23) and (1.24) *by inspection*. It is the authors' opinion that some preliminary inspection, with recognition of the answer as its goal, should precede the use of formal processes when treating almost any differential equation problem.

To complete the definition of the particular function $\phi(x)$ for which, as we demanded, $\phi(0) = 1$, we write, using Equation (1.24),

$$\phi(0) = \tfrac{1}{3}e^0 + Ae^0 = 1,$$

from which we obtain

$$A = \tfrac{2}{3}.$$

Thus, the final, uniquely determined, answer to our question is

$$\phi(x) = \tfrac{1}{3}e^{x^2} + \tfrac{2}{3}e^{-x^2/2}. \qquad (1.25)$$

Can you see that $\phi(x)$ is uniquely determined? Can you prove that it is? Sharpen up your technique with the following problems.

1.6 Problems

1.6.1 Find, for all x, that function $f(x)$ for which

$$f'(x) + xf(x) = x^{2m+1} \quad \text{with} \quad f(1) = 1.$$

The number m is any integer.

1.6.2 Solve, for all x with $f(0) = 0$,
$$f'(x) - 2xf(x) = 1.$$
Express the result in the form $f(x) = e^{x^2}g(x)$, where $g(x)$ is a definite integral (i.e., both limits are explicitly prescribed). Sketch an approximation to a graph of $g(x)$. Can you find $g(\infty)$ precisely?

1.6.3 Find the solution of
$$xh'(x) + Ah(x) = 1 + x^2 \quad \text{with} \quad h(1) = 1.$$
Over what x is the solution defined and continuous? Are there any values of the constant A for which the answer is peculiar? Ignore the condition $h(1) = 1$ and find that function $h(x)$ which obeys the differential equation and is bounded at the origin.

1.6.4 Find $u(x)$ such that, for *any* real number a,
$$u'(x) + au(x) = e^{-x} \quad \text{in} \quad 0 \le x < \infty$$
with $u(0) = 1$.

Sketch a graph of $u(x)$ for each of several values of a. Include, particularly, $a = -10^2, -1, 0, 1, 10^2, 10^4$.

1.6.5 Find $v(x)$ such that, for *any* real number ϵ,
$$\epsilon v'(x) + v(x) = 1 + x \quad \text{in} \quad 0 < x < \infty,$$
and $v(0) = 0$.

Sketch $v(x)$ for each of several values of ϵ; include, particularly, each of $\epsilon = 10^2, 1, 10^{-2}, 10^{-4}, 0, -10^{-2}, -1$.

In Problems 1.6.4 and 1.6.5, note that the behavior of $u(x)$ and $v(x)$ depends just as much on a and ϵ, respectively, as it does on x and that these functions might better have been called $u(x, a)$ and $v(x, \epsilon)$.

1.6.6 The rivers of this planet pour water into the oceans at the average rate of $4 \cdot 10^{19}$ grams per year (gm/yr). Necessarily, water evaporates from the oceans and is returned to the land at the same average rate. The average salinity of river water is 10^{-5} gm of salt per gm of water; the average salinity of the oceans is .035 gm of salt per gm of water; and salt is believed to accompany the evaporating ocean water (largely because of the evaporation of spray droplets) at such a rate that $3 \cdot 10^{14}$ gm of salt are returned to the land each year via rainfall. Accepting the dubious assumption that, over all geologic time, the salinity of river water and the evaporation rate are constant and that the salt return to the land is proportional to the salinity of ocean water, find and solve the differential equation whose solution describes the ocean salinity as a function of time. Draw any geophysical inferences you can from your result.

An alternative source of geophysical information estimates that the rivers provide the oceans with $3 \cdot 10^{15}$ gm of salt per year. How would this "fact" alter your geophysical inference?

1.7 A Nonlinear Equation

There are many problems in which the differential equation for, say, $u(x)$ is not linear but can be reduced to a linear equation by simple manipulations. For example, the differential equation

$$\theta'(x) + \sin[\theta(x)] = 0 \tag{1.26}$$

is a nonlinear equation for $\theta(x)$. However, we recall that $\theta'(x) = 1/x'(\theta)$ (everywhere?), and we note that Equation (1.26) can be recast in the form

$$x'(\theta) = -1/\sin\theta. \tag{1.27}$$

This is a linear equation for $x(\theta)$, the inverse of $\theta(x)$. The reader can verify that its general solution is

$$x = A - \ln\tan(\theta/2),$$

from which we obtain

$$\theta = 2\arctan(e^{A-x}). \tag{1.28}$$

An alternative technique is to use *separation of variables* and write Equation (1.26) in the equivalent form

$$\frac{d\theta}{\sin\theta} = -dx,$$

each side of which can be integrated so as to lead again to Equation (1.28).

1.8 Problems

1.8.1 When sugar is placed in a glass of iced tea and the tea is stirred vigorously, the prospective imbiber's thirst may intensify greatly before all the sugar passes into solution. If he recalls that water (and presumably tea) at $32°$ F can dissolve no more than 1.8 gm of sugar per cm^3 of water and that sugar passes into solution in pure water at about 10^{-4} cm^3 of sugar per sec per cm^2 of exposed surface, he can construct a mathematical model which might help him reach a reasonably quantitative understanding of the process. The plausible conjecture is that the rate at which sugar passes into solution in tea already containing sugar in solution is proportional to the exposed surface area but also proportional to $c_0 - c(t)$, where c_0 is the saturation concentration of sugar and c is the existing concentration. The proportionality must be such as to agree with the solution rate for pure water.

Find the appropriate differential equation and boundary conditions for the situation where m_0 gm of sugar for each cm^3 of tea [the sugar granules are to be spherical with initial diameter $\frac{1}{2}$ millimeter (mm)] are added to the tea at time zero. Find $m(t)$, the undissolved mass of sugar per cm^3 of tea, for various m_0; include especially values of m_0 for which $m_0 > c_0$ and some for which $m_0 < c_0$. Does proximity to saturation play an important role in the waiting time for iced-tea drinkers?

1.8.2   Rephrase the mothball problem of Section 1.3 so that it applies to the situation where many mothballs are enclosed in a container from which the vapor cannot escape. Make a plausible conjecture concerning the effect of the vapor concentration on the evaporation rate and solve the problem. Do you think the vapor concentration plays a significant role in any real situation?

1.8.3   Occasionally, a meteorite, an artificial satellite, or some other object enters the earth's atmosphere on a path approximately vertically downward. The force, F, which the atmosphere exerts on such a rapidly moving object is known to be adequately approximated by

$$F = \rho V^2 A C_D / 2,$$

where ρ is atmospheric density, V is the speed of the object, A is its projected cross-sectional area and C_D is a number which depends on the geometry; for blunt objects it can be taken as $C_D = 2$. If we confine our attention to objects that lose a negligible amount of their mass (by melting, evaporation, etc.) and approximate the atmospheric density distribution by

$$\rho = \rho_0 e^{-\beta y},$$

where y is altitude measured in cm, $\beta = 750{,}000$ cm^{-1}, and ρ_0 is sea level density, we can construct a mathematical model of the event and extract a reasonable description of the history, $V(t)$, of the object.

Find the appropriate differential equation for V^2 *as a function of y*, and determine the speed history of an object whose speed is observed to be V_0 at $y = 3 \times 10^7$ cm. Typical values of V_0 might be anything in the range 3×10^5 cm/sec $< V_0 < 3 \times 10^6$ cm/sec. Is the variation of the earth's gravitational pull with height important?

1.8.4 Find by inspection (that is, make an educated guess and check it) a solution of each of the following differential equations.

(a) $\quad u'(x) = x,$

(b) $\quad f'(x) + f(x) = e^x,$

(c) $\quad g'(x) + g(x) = 1,$

(d) $\quad h'(x) + (a/x)h(x) = 0,$

(e) $\quad w'(x) + (a/x)w(x) = x^3,$

(f) $\quad (xy'(x))' - x^{-1}y(x) = 0,$

(g) $\quad q''(x) + x^{-1}q'(x) - x^{-2}q(x) = 0,$

(h) $\quad a_n z^{(n)}(x) + a_{n-1} z^{(n-1)}(x) + \cdots + a_1 z'(x) + a_0 z(x) = 0,$

where each a_n is a constant and where $z^{(n)}$ denotes $d^n z/dx^n$. What features of each of the foregoing make it so simple that solution by inspection is the appropriate technique?

1.8.5 (a) An alternative way to obtain the general solution of Equation (1.17) is to try to find a function $g(x)$, an integrating factor, by which the equation may be multiplied so as to make the new left-hand side an exact derivative. In other words, we want

$$g(x)a_1(x)u' + g(x)a_0(x)u = [p(x)u]'$$

for some function $p(x)$. Follow this idea through to show that Equation (1.20) is again obtained, and observe that in essence this method is identical to that we have used.

(b) Equation (1.20) is not valid in any x-interval in which $a_1(x)$ vanishes; an example of this appeared in Problem 1.4.3. Let $a_1(x) = xa_2(x)$, where $a_2(0) \neq 0$, and discuss the nature of the solution (1.20) near $x = 0$. (Confine your attention to functions $a_0(x)$ and $a_1(x)$ for which $a_0'(0)$, $a_1'(0)$, $a_0''(0)$, $a_1''(0)$, ... all exist.)

First-Order Linear Difference Equations | 2

The differential equation (I.1) can be written in the form

$$\lim_{\alpha \to 0} \frac{f(x + \alpha) - f(x)}{\alpha} = \sigma f(x), \tag{2.1}$$

and it could be approximated (accurately, if α is small) by

$$\frac{f(x + \alpha) - f(x)}{\alpha} = \sigma f(x). \tag{2.2}$$

If we choose a fixed value for α in Equation (2.2), we can adopt the notation $x_n = n\alpha$ and write

$$f(x_{n+1}) - f(x_n) = \sigma \alpha f(x_n) \tag{2.3}$$

or, with $F_n = f(x_n)$,

$$F_{n+1} - F_n = \sigma \alpha F_n. \tag{2.4}$$

Equation (2.4) is a first-order linear homogeneous difference equation. Such equations arise directly in the treatment of scientific problems and are also encountered when differential equations are integrated by numerical means. It is of interest, then, to find those sets of numbers F_n which satisfy Equation (2.4) and to compare these numbers with appropriate values of the function $f(x)$ associated with Equation (I.1).

Provided that $F_0 \neq 0$, we can restrict our attention *without loss of generality* (the reader should verify the truth of this italicized remark) to solutions of Equation (2.4) for which $F_0 = 1$.

Using $F_0 = 1$ and Equation (2.4), we obtain
$$F_1 = (1 + \sigma\alpha), \quad F_2 = (1 + \sigma\alpha)F_1 = (1 + \sigma\alpha)^2,$$
and, generally,
$$F_n = (1 + \sigma\alpha)^n. \tag{2.5}$$
This can be written as
$$F_n = e^{n\ln(1+\sigma\alpha)}. \tag{2.6}$$
The corresponding solution to Equation (I.1) (i.e., that for which $f(0) = 1$) is
$$f(x) = e^{\sigma x},$$
and, in particular, for points $x = n\alpha$,
$$f(n\alpha) = e^{n\sigma\alpha}. \tag{2.7}$$
The reader can and should make the appropriate comparison.

The general first-order linear homogeneous difference equation has the form
$$G_{n+1} = a_n G_n, \tag{2.8}$$
where the a_n are given coefficients.*

The additional constraint, $G_{n_0} = A$, allows us to write
$$G_{n_0+1} = A a_{n_0}, \quad G_{n_0+2} = a_{n_0+1} G_{n_0+1} = A a_{n_0} a_{n_0+1}$$
and, in general,
$$G_N = A \prod_{j=n_0}^{N-1} a_j = A \exp \sum_{j=n_0}^{N-1} \ln a_j. \tag{2.9}$$
(The symbol $\Pi_{j=n}^{m} a_j$ means the product of the numbers a_n, a_{n+1}, ..., a_m.)

It is clear from the elementary construction leading to Equation (2.9) that only one constraint (boundary condition) of the form $G_{n_0} = A$ is required to imply a unique solution for a first-order linear homogeneous difference equation.

The general solution of the first-order linear nonhomogeneous difference equation
$$F_{n+1} - a_n F_n = c_n \tag{2.10}$$
with
$$F_{n_0} = B$$

---

* One could insist on the general form $b_n G_{n+1} = c_n G_n$ but, unless some of the b_n vanish, the definition $a_n = c_n/b_n$ can be made immediately. The reader can easily infer the implications of $b_n = 0$.

can be constructed, using a scheme analogous to that leading to Equation (1.21). We define

$$\alpha_{n+1} = \prod_{j=n_0}^{n} a_j.$$

For any n for which $\alpha_{n+1} \neq 0$, Equation (2.10) can be written as

$$\frac{F_{n+1}}{\alpha_{n+1}} - \frac{F_n}{\alpha_n} = \frac{c_n}{\alpha_{n+1}}, \qquad (2.11)$$

and, with the abbreviations $F_n/\alpha_n = H_n$ and $p_n = c_n/\alpha_{n+1}$, we have

$$H_{n+1} - H_n = p_n. \qquad (2.12)$$

Since we know that $H_{n_0} = B/\alpha_{n_0}$, we find that

$$H_{n_0+1} = B/\alpha_{n_0} + p_{n_0}, \qquad (2.13)$$

$$H_{n_0+2} = p_{n_0+1} + H_{n_0+1} = B/\alpha_{n_0} + p_{n_0} + p_{n_0+1},$$

and, in general,

$$H_n = B/\alpha_{n_0} + \sum_{j=n_0}^{n-1} p_j. \qquad (2.14)$$

Finally, replacing the abbreviations, we have

$$F_n = B \frac{\alpha_n}{\alpha_{n_0}} + \alpha_n \sum_{j=n_0}^{n-1} \frac{c_j}{\alpha_{j+1}}. \qquad (2.15)$$

2.1 Problems

2.1.1 Find the solution of

$$G_{n+1} - nG_n = 1 - n,$$

with $G_1 = 2$. For which integers n is G_n defined?

2.1.2 Find the general solution of

$$(n + 1)F_{n+1} - nF_n = a^n.$$

What is F_n when $F_1 = 1$? For what value of F_1 (if any) does F_n exist for each integer n and satisfy the given difference equation? Find these F_n. Here $a > 0$ is a constant.

2.1.3 Radium transmutes spontaneously at such a rate that, after 25 years, only .99 of any piece of radium remains unchanged. Find the difference equation which is equivalent to this fact. Solve this equation and infer $m(t)$, the mass of radium which remains at time t if, at time zero, there were m_0 gm of radium. Rephrase this problem in such a way that the given information would lead directly to a differential equation with the same solution at appropriate times.

2.1.4 Write a difference equation approximation to the differential equation

$$[x\phi(x)]' = e^{\alpha x}.$$

Compare it with the difference equation of Problem 2.1.2, and compare its general solution with that found in this Problem. In particular, identify the conditions under which the differential equation may be used to find a suitable approximation to F_n.

2.1.5 As in the case of differential equations, it is often worthwhile to try to guess the solution of a difference equation before starting to apply a general formula. For example, the reader may have noticed that Problems 2.1.1 and 2.1.2 become very easy if we make the transformations $G_n = 1 + H_n$, $nF_n = H_n$, respectively. Similarly, find a simple transformation that reduces the second-order difference equation problem

$$F_{n+1} - 2F_n + F_{n-1} = n$$

to a sequence of first-order problems, and thus obtain its general solution.

2.1.6 In a tight-money period, interest rates gradually rise. Assuming that, as a first approximation, the interest rate paid semi-annually on savings deposits increases linearly with time, determine the accumulated value after n interest periods of a regular semi-annual deposit of P dollars.

Second-Order Differential Equations | 3

Each second-order linear ordinary differential equation has the form

$$a_2(x)u''(x) + a_1(x)u'(x) + a_0(x)u(x) = g(x). \qquad (3.1)$$

There are many alternative forms into which this can be cast. For example, in any x-interval in which $a_2(x)$ does not vanish, the reader should verify that

$$[p(x)u'(x)]' + q(x)u(x) = h(x) \qquad (3.2)$$

is precisely the same statement as Equation (3.1), provided that

$$\ln p(x) = \int^x [a_1(x)/a_2(x)]\, dx, \qquad q(x) = a_0(x)p(x)/a_2(x),$$

and

$$h(x) = g(x)\, p(x)/a_2(x).$$

Furthermore, Equation (3.1) is implied by the following pair of first-order equations:

$$a_2(x)v'(x) + a_1(x)v(x) + a_0(x)u(x) = g(x) \qquad (3.3)$$

and

$$u'(x) - v(x) = 0. \qquad (3.4)$$

Finally, it is sometimes convenient to "suppress the first derivative," i.e., to write Equation (3.1) in the form

$$w''(x) + r(x)w(x) = m(x). \qquad (3.5)$$

The reader should show that this can be accomplished via the transformations

$$w = u(x)[p(x)]^{1/2} = u(x) \exp\left[\int^x \frac{a_1(t)}{2a_2(t)} dt\right],$$

$$r = \frac{a_0}{a_2} - \frac{a_1^2}{4a_2^2} - \frac{1}{2}\left(\frac{a_1}{a_2}\right)', \quad \text{and} \quad m = \frac{g(x)[p(x)]^{1/2}}{a_2(x)}.$$

Not only can Equation (3.1) be cast in the form of Equation (3.2) or Equations (3.3) and (3.4) or Equation (3.5) but, also, an equation of the form (3.2) can be cast in the form of Equation (3.1), and any pair of ordinary first-order differential equations that are linear in $u(x)$, $u'(x)$, $v(x)$, and $v'(x)$ can be cast in the form of Equation (3.1). We leave to the reader the manipulations that establish these facts. Thus, any results we derive by using Equations (3.1), (3.2), (3.3) and (3.4) or (3.5) can be carried over to each of the others.

The most frequently encountered second-order equation is that with constant coefficients. Let a and b be constants, and consider

$$u'' + au' + bu = g(x), \tag{3.6}$$

which can be written in the form of Equation (3.5) as

$$w'' + \alpha w = g(x)e^{ax/2}, \tag{3.7}$$

where

$$w = ue^{ax/2} \quad \text{and} \quad \alpha = b - a^2/4.$$

It is advantageous to consider first the homogeneous equation corresponding to Equation (3.7), that is,

$$w'' + \alpha w = 0. \tag{3.8}$$

Drawing again on the knowledge gained in the study of the calculus, we recall that

$$\frac{d^2}{dx^2}(\sin \beta x) = -\beta^2 \sin \beta x$$

and that

$$\frac{d^2}{dx^2}(\cos \beta x) = -\beta^2 \cos \beta x.$$

It follows immediately that, when $\alpha > 0$, any multiple of $\sin(x\sqrt{\alpha})$ or of $\cos(x\sqrt{\alpha})$ is a solution of the homogeneous equation. When $\alpha < 0$, we need merely recall that the derivatives of the exponential function are such that any multiple of $e^{x\sqrt{-\alpha}}$ or $e^{-x\sqrt{-\alpha}}$ is a solution of Equation (3.8).

We do not really need to make this distinction between the two cases $\alpha > 0$ and $\alpha < 0$. For if c is any complex constant of the form $c = a + ib$, where a and b are real constants, then from rudimentary complex variable theory we obtain

$$e^{cx} = e^{ax}(\cos bx + i \sin bx),$$

from which it follows that

$$\frac{d}{dx}(e^{cx}) = ce^{cx}.$$

This is the same as the usual formula for the case in which c is real. Thus,

$$e^{ix\sqrt{\alpha}}, \qquad e^{-ix\sqrt{\alpha}}$$

are solutions of Equation (3.8), irrespective of the sign of α; more generally, any linear combination $Ae^{ix\sqrt{\alpha}} + Be^{-ix\sqrt{\alpha}}$, where A and B are any complex constants, is a solution. Choosing $A = B = \frac{1}{2}$ or $A = -B = \frac{1}{2}$, we see that $\cosh(ix\sqrt{\alpha})$ and $\sinh(ix\sqrt{\alpha})$ are also solutions, as is any linear combination of them. The choices $A = B = \frac{1}{2}$ and $A = -B = 1/2i$ lead to the solutions $\cos(x\sqrt{\alpha})$ and $\sin(x\sqrt{\alpha})$. Apparently we have a variety of possible descriptions of the solutions of Equation (3.8), and there may be others we have not listed.

At this point the reader should

(1) identify solutions of Equation (3.8) for $\alpha = 0$;
(2) ascertain that each of the functions $\sin(x\sqrt{\alpha})$, $\cos(x\sqrt{\alpha})$, $\sinh(ix\sqrt{\alpha})$, and $\cosh(ix\sqrt{\alpha})$ can be written as a linear combination of $e^{ix\sqrt{\alpha}}$ and $e^{-ix\sqrt{\alpha}}$;
(3) ascertain that any of the functions $\sin(x\sqrt{\alpha}), \cos(x\sqrt{\alpha}), e^{ix\sqrt{\alpha}}$, and $e^{-ix\sqrt{\alpha}}$ can be written as a linear combination of *any* pair of the remaining three functions;
(4) begin to suspect that *there are only two linearly independent* solutions of Equation (3.8).*

To show that the suspicion of item (4) is indeed well-founded, we note that Equation (3.8) can be written in the form

---

* N functions, $f_n(x)$, are said to be linearly dependent in the interval $a < x < b$ if there exist constants, c_n, of which at least two are different from zero, such that $\sum c_n f_n(x) = 0$ at every x in the interval. All this says is that if N functions are linearly independent, no one of them can be equated to a linear combination of the others.

$$\left(\frac{d}{dx} - i\sqrt{\alpha}\right)\left(\frac{d}{dx} + i\sqrt{\alpha}\right) w = 0, \tag{3.9}$$

and we can define

$$H = \left(\frac{d}{dx} + i\sqrt{\alpha}\right) w, \tag{3.10}$$

so that, by Equation (3.9), we obtain

$$\left(\frac{d}{dx} - i\sqrt{\alpha}\right) H = 0. \tag{3.11}$$

We know from Chapter 1 that the most general solution of Equation (3.11) is

$$H = Ke^{ix\sqrt{\alpha}},$$

where K is a (complex) constant. With H known, Equation (3.10) is a differential equation for w; its most general solution we know (again from Chapter 1) to be

$$w = De^{-ix\sqrt{\alpha}} + (K/2i\sqrt{\alpha})e^{ix\sqrt{\alpha}} \tag{3.12}$$

(where D is a constant) unless $\alpha = 0$, in which case

$$w = Mx + N,$$

where M and N are constants.

But if Equation (3.12) represents the most general solution of Equation (3.9), then any solution at all must have this form and be a linear combination of the two solutions $e^{-ix\sqrt{\alpha}}$ and $e^{ix\sqrt{\alpha}}$. Therefore, there are indeed two, and not more than two, linearly independent solutions of Equation (3.8). Any of the functions listed under (3) above may be adopted as the fundamental pair, and the investigator of any problem should choose such a fundamental pair to suit his own convenience.

The homogeneous equation corresponding to Equation (3.6) is

$$u'' + au' + bu = 0,$$

and it now follows that the most general solution of this equation has the form

$$u = Ke^{r_1 x} + De^{r_2 x}, \tag{3.13}$$

where K and D are as yet undetermined constants and where r_1 and r_2 are the two roots of

$$r^2 + ar + b = 0.$$

This statement must be modified slightly if r_1 and r_2 are equal (i.e., if $a^2 = 4b$); the most general solution is then

$$u = Ke^{r_1 x} + Dxe^{r_1 x}.$$

Referring to Equation (3.6), suppose we can find (by guessing or by techniques to be discussed later) *any* function $h(x)$ satisfying the equation. Then the function $u - h$ satisfies the homogeneous counterpart of the equation and, hence, must have the form of the right-hand side of Equation (3.13); thus the most general solution of Equation (3.6) is

$$u = h(x) + Ke^{r_1 x} + De^{r_2 x}$$

or the minor modification of this if $r_1 = r_2$. The constants K and D may be chosen so as to satisfy whatever added conditions are imposed on u. For example, to solve the equation

$$au'' + bu' + cu = Ae^{mx}, \qquad (3.14)$$

where a, b, c, A, m are constants, we first observe that a suitable choice for $h(x)$ is Ke^{mx}, where K is a constant to be determined. Thus, we write

$$u = Ke^{mx} + Be^{\nu x} + Ce^{\mu x},$$

where K, ν, μ are to be found and where B and C can be any numbers (unless further constraints on $u(x)$ are imposed). Substitution yields

$$[K(am^2 + bm + c) - A]e^{mx} + C(a\mu^2 + b\mu + c)e^{\mu x}$$
$$+ B(a\nu^2 + b\nu + c)e^{\nu x} = 0,$$

so that

$$K = A/(am^2 + bm + c)$$

and ν and μ are the roots of the quadratic equation

$$as^2 + bs + c = 0.$$

Our solution is invalid, of course, for either of two particular values of m; what are they? Can you find $u(x)$ when m has one of those values?

In the above solution, the constants B and C are as yet undetermined. Usually, certain constraints on $u(x)$ will be imposed, and these will suffice to determine B and C. For example, the values of $u(0)$ and $u'(0)$ might be prescribed; alternatively, $u(0)$ and $u(1)$ could be given, and other such constraints, often called *boundary conditions* or *initial conditions*, are also possible. Ordinarily, in problems arising from scientific and engineering contexts, the constraints will be of the form

$$K_{11}u(x_1) + K_{12}u'(x_1) = M_1, \qquad K_{21}u(x_2) + K_{22}u'(x_2) = M_2, \qquad (3.15)$$

where the K_{ij}, M_i, and x_i are given numbers. When $x_1 \neq x_2$, the range of x over which a description of $u(x)$ is wanted will ordinarily be the interval $x_1 < x < x_2$. When $x_1 = x_2$, the typical range will be $x > x_1$ or $x < x_1$.

Sometimes it is useful to choose the fundamental pair of solutions of the homogeneous counterpart of the equation in a way that will simplify the application of these constraints. For example, suppose that we are interested in those solutions of

$$u'' + \lambda u = e^{kx}$$

(with $\lambda > 0$) for which $u(0) = R$. These solutions are most neatly studied by using $\sin x\sqrt{\lambda}$ and $\cos x\sqrt{\lambda}$ as the fundamental pair of solutions of

$$w'' + \lambda w = 0$$

and writing

$$u = Le^{kx} + A \sin x\sqrt{\lambda} + B \cos x\sqrt{\lambda}.$$

The constant L is found by substitution into the differential equation; B can then be found at once by the condition $u(0) = R$, and A remains as the single free parameter. Thus, the functions sought have the form

$$u = \frac{1}{k^2 + \lambda} e^{kx} + \left(R - \frac{1}{k^2 + \lambda}\right) \cos(x\sqrt{\lambda}) + A \sin x\sqrt{\lambda},$$

where A is any constant.

Another useful observation states that, when

$$u'' + k^2 u = 0$$

and $u(x_0) = R$, a useful choice (why?) for the fundamental pair of solutions is

$$u_1 = \sin[k(x - x_0)] \quad \text{and} \quad u_2 = \cos[k(x - x_0)].$$

3.1 Problems

3.1.1 For each real number A (no exceptions), let

$$u''(x) + 2u'(x) + Au(x) = 0.$$

(a) Find $u(x)$ in $0 \leq x < \infty$ such that $u(0) = 0$, $u'(0) = 1$. Is the solution uniquely determined?
(b) Find $u(x)$ in $0 \leq x \leq 1$ such that $u(0) = 0$, $u'(1) = 1$.

(c) Find $u(x)$ in $0 \leq x \leq L$ such that $u(0) = 0$, $u'(L) = 0$, where $L = (1 - A)^{-1/2}$.

3.1.2 Let A be a real constant. For
$$u''(x) + A^2 u(x) = g(x):$$
(a) let $g(x) = 1$. Find all functions $u(x)$ such that $u(0) = 0$. Find all functions $u(x)$ such that $u(0) = u(1) = 0$. Find all functions $u(x)$ such that $u(0) = 0$ and $u(1) = 1$.
(b) Let $g(x) = \sin x$. Find all functions $u(x)$ obeying the requirements of part (a).
(c) Repeat (b) with $g(x) = \sin 2x$, with $g(x) = \sin kx$; in each case, what value of A is rather special?
(d) Repeat (b) with $g(x) = e^{kx}$.

3.1.3 The population, $P(t)$, of a certain kind of parasite in a human host increases at a rate proportional to the population if the temperature of the host remains normal. However, this rate of increase is lessened by an amount proportional to $Y(t) = T(t) - T_0$, where T is the host's temperature and T_0 is his normal temperature. The host responds to the parasite according to the rule
$$Y' + \alpha Y = kP.$$
Find the equation governing the population growth of the parasite; choose initial conditions such that the patient is healthy until, at time $t = 0$, he is mildly infected with the parasite; solve the differential equations subject to these initial conditions. What relations (inequalities) among α, k, and the other parameters of the problem imply the following prognoses: (a) death, (b) survival after a crisis, or (c) negligible illness?

3.1.4 A mathematically-minded vandal wishes to break a steam radiator away from its foundation, but finds that when he applies steadily the greatest force of which he is capable [100 kilograms (kg)] the 2-cm displacement of the top of the radiator is only one-tenth of that required for his purposes. He finds, however, that he can apply a force $f(t)$ according to the schedule
$$f(t) = (F/2)(1 - \cos \omega t), \quad \text{where} \quad F = 100 \text{ kg}$$
for any of a large range of values of ω. The mass of the radiator is 50 kg and its foundation resists its movement by a force proportional to its displacement. At what frequency and for how long must he exert the force $f(t)$ if he is to succeed?

3.1.5 Let $\epsilon u''(x) - u(x) = 1$ in $|x| < 1$ with $u(-1) = u(1) = 0$. Find $u(x)$ and sketch its graph for $\epsilon = 10^{-4}$, 10^{-2}, 1, and 0. Review what you learned from Problems 1.6.4 and 1.6.5. Can you infer the nature of the role played by ϵ to estimate, for $\epsilon \ll 1$, without exact analytical descriptions of $w(x)$ and $y(x)$, the solutions of

$$\epsilon w'(x) + (1 + x^2)w(x) = 1 \quad \text{in each of} \quad x < 0, x > 0 \text{ with } w(0) = 0,$$

and

$$\epsilon y''(x) - (1 + x^2)y(x) = 1 \quad \text{in} \quad -1 < x < 1$$
$$\text{with} \quad y(-1) = y(1) = 0?$$

This one is difficult.

3.1.6 A buoy having the shape of a circular cylinder floats in the ocean with its axis vertical; in smooth water, $\frac{3}{4}$ of its length is submerged. Using the fact that the buoyant force is equal to the weight of displaced water, determine the steady-state motion of the buoy resulting from the passage of sinusoidal waves, of wave length λ, past the buoy. Assume that the effect of the mooring chain is negligible, that λ is large compared to the diameter of the buoy, and that the motion of the buoy is purely vertical. Are there any special cases of interest? What effect would the mooring chain have?

3.1.7 A crystal element in a satellite radio transmitter is to be maintained at (nearly) constant temperature by means of a network of small heating elements placed close to the crystal. The crystal, whose average temperature T is proportional to its heat content, loses heat to a surrounding container (average temperature T_c; heat content proportional to T_c) at a rate proportional to $T - T_c$. In turn, the container loses heat to the external environment (temperature T_0) at a rate proportional to $T_c - T_0$. The temperature T_0 will usually fluctuate with time. A control circuit monitors the temperature T_c, and provides a heating current to the crystal. It is proposed that the rate of heat input to the crystal, $f(t)$, be a linear function of T_c and dT_c/dt. How would you choose the coefficients of this linear combination? Is such a control system effective?

3.1.8 A solution of the homogeneous counterpart of Equation (3.6) is $e^{r_1 x}$, where r_1 is one root of $r^2 + ar + b = 0$. Show that if we define a new function w by $u = e^{r_1 x} w$, and substitute for u in Equation (3.6), the function $w(x)$ can be determined in terms of integrals involving $g(x)$. How many undetermined constants appear in the resulting solution $u(x)$?

3.2 The Homogeneous Problem

There is an exceptionally important class of mathematical problems of which Problem 3.1.1(c) is an atypical example. A *typical* homogeneous problem asks: what functions $u(x)$ are solutions of

$$u'' + \lambda u = 0 \quad \text{in} \quad 0 < x < 2 \tag{3.16}$$

with

$$u(0) = u(2) = 0? \tag{3.17}$$

The most general solution, u, of Equation (3.16) is

$$u = A \sin(x\sqrt{\lambda}) + B \cos(x\sqrt{\lambda}),$$

where A and B are constants. In order that $u(0) = 0$, B must be zero. If $u(2)$ is to be zero, either A must vanish *or* $\sin(2\sqrt{\lambda})$ must be zero. Thus as we could have seen immediately, $u(x) = 0$ satisfies the requirements of our problem for any value of λ and is called a *trivial* solution of the problem. However, when λ is any of the numbers,

$$\lambda_n = n^2 \frac{\pi^2}{4}, \quad n = 1, 2, 3, \ldots,$$

any multiple of

$$u_n(x) = \sin(x\sqrt{\lambda_n}) = \sin\left(\frac{n\pi x}{2}\right)$$

satisfies those requirements. *We conclude that only for a certain countable* set of values of λ is there a nontrivial solution of this typical homogeneous problem.* These values of λ (that is, $\lambda_n = (n\pi/2)^2$) are called the eigenvalues of the problem and the corresponding solutions $u_n(x)$ are called its eigenfunctions. Eigenvalue problems which arise in connection with second order differential equations ordinarily are characterized by

(1) a homogeneous differential equation which can be cast in the form

$$[p(x)u'(x)]' + q(x)u(x) + \lambda h(x)u(x) = 0 \quad \text{in} \quad x_1 < x < x_2;$$

---

* An infinite set of quantities is said to be *countable* if the members can be put into a one-to-one correspondence with the natural numbers 0, 1, 2, 3, ... Thus, since the list $\frac{1}{1}, \frac{1}{2}, \frac{1}{3}, \frac{2}{3}, \frac{1}{4}, \frac{2}{4}, \frac{3}{4}, \ldots$ includes all the rational numbers, the set of rational numbers in [0,1] is countable.

(2) a pair of homogeneous boundary conditions

$$k_{11}u(x_1) + k_{12}u'(x_1) = 0, \qquad k_{21}u(x_2) + k_{22}u'(x_2) = 0;$$

(3) a question of the following type: for what values of λ do nontrivial functions $u(x)$ satisfy these equations? For each of these values of λ, what are the functions $u(x)$?

The solutions of eigenvalue problems have many useful and interesting properties to which we shall return later. In the meantime, the reader should carry out the following exercises.

3.3 Problems

3.3.1 Show that, for the problem of Equations (3.16) and (3.17)

$$\int_0^2 u_n(x)u_m(x)\,dx = 0 \qquad \text{for} \quad n \neq m.$$

3.3.2 What function(s) $v_m(x)$ and number(s) C_m satisfy the following requirements

$$v_m''(x) + C_m v_m(x) = 0 \qquad \text{in} \quad 3 < x < 4$$

and

$$v_m(3) = v_m(4) = 0\,?$$

3.3.3 What functions $y_n(x)$ and numbers B_n satisfy the requirements

$$y_n''(x) + B_n y_n(x) = 0 \qquad \text{in} \quad 0 < x < \pi,$$

and

$$y_n(0) = 0, \qquad y_n'(\pi) - y_n(\pi) = 0?$$

3.3.4 Find the eigenvalues, A_m, and the eigenfunctions, $u_m(x)$, of the homogeneous problem

$$u''(x) + 2u'(x) + Au(x) = 0 \qquad \text{in} \quad 0 < x < 1 \qquad (3.18)$$

with

$$u(0) = u'(1) = 0. \qquad (3.19)$$

3.3.5 Review Problem 3.1.1 and compare any peculiarities with the implications of Problem 3.3.4. Does any general conjecture occur to you? Test the conjecture by inventing an appropriate pair of problems and comparing their solutions.

3.3.6 Find the eigenvalues and eigenfunctions of

$$u''(x) + \lambda p(x)u(x) = 0 \quad \text{in } 0 < x < 1 \quad (3.20)$$

with

$$u(0) = u(1) = 0 \quad (3.21)$$

and with

$$p(x) = \begin{cases} 1, & x < \tfrac{1}{2}, \\ 4, & x > \tfrac{1}{2}. \end{cases} \quad (3.22)$$

In this problem we mean by an eigenfunction a function with as many continuous derivatives as we can demand without eliminating everything but $u(x) \equiv 0$. Thus in this problem, $u_n(x)$ and $u'_n(x)$ should be continuous; can you see that requiring continuity of $u''(x)$ is prohibitively restrictive?

3.3.7 Study the eigenfunctions of Equations (3.16) and (3.17), and those of Problems 3.3.4 and 3.3.6, sketching graphs of many of them. What features do the three different sets of eigenfunctions have in common?

3.4 Operators Which Can Be Factored

Occasionally, one encounters a second-order differential equation

$$\left(a_2(x) \frac{d^2}{dx^2} + a_1(x) \frac{d}{dx} + a_0(x) \right) u(x) = g(x) \quad (3.23)$$

which can be written in the form

$$\left[\alpha_1(x) \frac{d}{dx} + \beta_1(x) \right]\left[\alpha_2(x) \frac{d}{dx} + \beta_2(x) \right] u(x) = g(x). \quad (3.24)$$

When this is the case, the second-order differential operator,

$$a_2 \frac{d^2}{dx^2} + a_1 \frac{d}{dx} + a_0,$$

is said to have factors; namely, the first order operators,

$$L_1 = \left(\alpha_1 \frac{d}{dx} + \beta_1 \right) \quad \text{and} \quad L_2 = \left(\alpha_2 \frac{d}{dx} + \beta_2 \right).$$

Defining

$$m(x) = \alpha_2(x)u'(x) + \beta_2(x)u(x), \quad (3.25)$$

the reader can verify that those $u(x)$ which satisfy Equation (3.23) can be found by solving Equation (3.24) as a first-order equation for $m(x)$ and then solving Equation (3.25) as a first-order equation for $u(x)$.

3.5 Problems

3.5.1 Find all functions, $u(x)$, such that
$$xu'' - (1 + x)u' + u = 0.$$

3.5.2 Find all functions, $f(x)$, such that
$$x^2 f'' + xf' - f = 0 \quad \text{in} \quad x > 1,$$
with
$$f(1) = 0, \quad f'(1) = A.$$

3.5.3 Find those $w(x)$ for which
$$w''(x) + xw'(x) = 0 \quad \text{in} \quad 0 < x < \infty$$
with
$$w(0) = 1, \quad w'(0) = 1.$$

3.5.4 Over any interval on which the function $a_2(x)$ of Equation (3.23) does not vanish, we can divide Equation (3.23) through by $a_2(x)$ so as to make the coefficient of the leading term unity. The differential operator then takes the form
$$L = \frac{d^2}{dx^2} + a_1(x)\frac{d}{dx} + a_0(x).$$

Show that if $u = y(x)$ is any solution of $Lu = 0$, then L can be factored via
$$L \equiv \left(\frac{d}{dx} + \left[a_1(x) + \frac{y'}{y}\right]\right)\left(\frac{d}{dx} - \frac{y'}{y}\right).$$

Deduce that the general solution to a second-order differential equation can involve at most two arbitrary constants.

3.5.5 Show that the transformation $y' = wy$ changes the general second-order linear homogeneous differential equation
$$a_2(x)y'' + a_1(x)y' + a_0(x)y = 0$$
into a first-order nonlinear equation.

3.5.6 Construct a mathematical problem which models both the loss of heat from a home to the outside and the supply of heat from the automatically controlled furnace to the home. Let the heating system be a two-zone system. Take account of the heat transfer between the two zones (say A and B). Find $T_A(t)$ and $T_B(t)$ under initial conditions of your own choosing. Find $T_A(t)$ and $T_B(t)$ under the same initial conditions when a mistake has been made in the wiring, so that the thermostat for zone A has been connected to the control for zone B, and vice versa.

Power Series Descriptions | 4

Ordinarily, when one is required to solve a differential equation, one wants to obtain a description of the solution which can be easily interpreted in the context of the question from which the equation arose. Thus far, both in the text and in the Problems, ease of interpretation has been assured because the solution of each equation has been described in terms of a simple combination of elementary functions. Since an elementary function is nothing more special than a function which is so familiar to the one who encounters it that he can recall or extract easily any of its properties he needs,* such a description is greatly to be desired. Unfortunately, however, no matter how large one's "vocabulary" of elementary functions may be, the solutions of most differential equations cannot be written as simple combinations of a finite number of elementary functions; accordingly, we must find systematic procedures by which we can construct solutions to a broad class

---

* For example, the trigonometric function, $\sin x$, is an elementary function from the reader's point of view because he recalls that $\sin x$ is that odd, oscillatory, smooth function for which $|\sin x| \leq 1$, $\sin x = \sin(x + 2\pi)$ and for which meticulous numerical information can be found in easily accessible tables; the Bessel function, $J_0(x)$, on the other hand, probably won't be an elementary function to that same reader until he has thoroughly digested Chapter 11. Then, $J_0(x)$ will be that familiar, even, oscillatory, smooth function which, when $x \gg 1$, is closely approximated by $\sqrt{2/\pi x} \cos(x - \pi/4)$, which obeys the differential equation, $(xu')' + xu = 0$, for which $J_0(0) = 1$, and for which meticulous numerical information can be found in easily accessible tables.

of differential equations. We can expect that, in general, such procedures will lead to clumsier descriptions than we have seen in the foregoing and that several different representations of a given function may be needed to describe it conveniently over a large range of the independent variable.

One broadly applicable procedure is that which provides the power series representation of a solution of a differential equation. We initiate our study of power series representations by asking: does the equation

$$u''(x) + xu(x) = 0 \qquad (4.1)$$

have a solution of the form

$$u(x) = \sum_{n=0}^{\infty} a_n x^n \qquad (4.2)$$

which converges in some x-interval?

An obvious approach is to try to determine the coefficients a_n of any such solution by substituting Equation (4.2) into Equation (4.1) and "collecting" coefficients of each power of x. In doing all this, we differentiate term by term and regroup terms with optimism but without formal justification. We obtain

$$2a_2 + (6a_3 + a_0)x + (12a_4 + a_1)x^2 + \cdots$$
$$+ [n(n-1)a_n + a_{n-3}]x^{n-2} + \cdots = 0. \qquad (4.3)$$

If the series on the left-hand side of Equation (4.3) is to converge to zero for all small enough x, the coefficient of each power of x must vanish. Therefore

$$a_2 = 0,$$
$$a_3 = -a_0/(2 \cdot 3),$$
$$a_4 = -a_1/(3 \cdot 4),$$
$$a_5 = -a_2/(4 \cdot 5) = 0,$$
$$a_6 = -a_3/(5 \cdot 6) = a_0/(2 \cdot 3 \cdot 5 \cdot 6),$$
$$\vdots$$

More generally, we have

$$a_{3n} = -\frac{a_{3(n-1)}}{3n(3n-1)} = \frac{a_{3(n-2)}}{3n(3n-1) \cdot 3(n-1)[3(n-1)-1]}$$
$$= \cdots$$
$$= \frac{(-1)^n a_0}{3^{2n} n!(n - \frac{1}{3})(n - \frac{4}{3}) \cdots (\frac{2}{3})},$$

$$a_{3n+1} = -\frac{a_{3n-2}}{(3n+1)(3n)} = \frac{a_{3n-5}}{(3n+1)(3n)(3n-2)(3n-3)}$$

$$= \cdots$$

$$= \frac{(-1)^n a_1}{3^{2n} n!(n+\frac{1}{3})(n-\frac{2}{3})\cdots(\frac{4}{3})},$$

and

$$a_{3n+2} = 0.$$

If we denote by $u_1(x)$ and $u_2(x)$ the functions

$$u_1(x) = \sum_{n=0}^{\infty} a_{3n} x^{3n} \tag{4.4}$$

and

$$u_2(x) = \sum_{n=0}^{\infty} a_{3n+1} x^{3n+1}, \tag{4.5}$$

we note immediately that

(1) any power series that can be constructed by the foregoing process leads to a linear combination of $u_1(x)$ and $u_2(x)$;

(2) $u_1(x)$ and $u_2(x)$ are linearly independent (note that a nontrivial function which vanishes at the origin cannot be a multiple of one that does not);

(3) the power series for each of $u_1(x)$ and $u_2(x)$ converges absolutely and uniformly in any region $|x| \leq m$, where m is any positive number;

(4) it would be rather difficult, without some sophisticated technique, to infer from the series representation of $u_1(x)$ and $u_2(x)$ any clear quantitative picture of the manner in which $u_1(x)$ and $u_2(x)$ change with x as x becomes large.

We will frequently encounter the functions $u_1(x)$ and $u_2(x)$ of Equations (4.4) and (4.5); hopefully, they will soon become familiar and useful. In the meantime, further insight into the use of power series can be gained by carrying out the following exercises.

4.1 Problems

4.1.1 *Prove* that $u_1(x)$ and $u_2(x)$ as given by Equations (4.4) and (4.5) are solutions of Equation (4.1).

4.1.2 How many terms of the series representation of $u_1(x)$ would be needed to calculate $u_1(1{,}000)$ with an error E for which $|E| < 10^{-3}$?

4.1.3 Do Problems 3.1.1, 3.1.5, 3.3.4, 3.5.1, 3.5.2, 3.5.3 again, using power series descriptions of the solutions of the differential equations. Carry each problem only to the point where

(a) you see that it cannot be done that way,
(b) you see that the process is too cumbersome to give an easily interpretable answer, or
(c) you can plot a graph of the solutions.

For the problems in category (c), have you really used power series representations, polynomial approximations, or both?

4.2 A More Recalcitrant Illustrative Problem

Another illustrative problem has a rather different result. We let

$$x^2 w''(x) - (2 + x^2)w = 0 \qquad (4.6)$$

and ask: Can any function $w(x)$ which satisfies Equation (4.6) be described in the form

$$w(x) = \sum_{n=0}^{\infty} a_n x^n \;? \qquad (4.7)$$

We substitute Equation (4.7) into Equation (4.6) as before, and we find

$$-2a_0 - 2a_1 x - a_0 x^2 + (4a_3 - a_1)x^3 + (10a_4 - a_2)x^4 + \cdots \\ + \{[n(n-1) - 2]a_n - a_{n-2}\}x^n + \cdots = 0, \qquad (4.8)$$

so that

$$a_0 = 0,$$
$$a_1 = 0,$$
$$a_3 = a_1/4 = 0,$$
$$a_4 = a_2/10,$$
$$\vdots$$
$$a_n = a_{n-2}/(n-2)(n+1) \qquad \text{for } n > 2.$$

By the method used in the foregoing problem we can deduce from this difference equation that, for $n \geq 2$,

§4.2] A More Recalcitrant Illustrative Problem 37

$$a_{2n} = a_2 \frac{3(2n)}{(2n+1)!},\qquad(4.9)$$

$$a_{2n+1} = 0.$$

Thus, once we have chosen a_2, the series is uniquely determined and there is only one solution of Equation (4.6) which has a power series about $x = 0$ and which is linearly independent of all other such solutions. We shall discuss this particular situation in more detail, but first we ask whether a simple extension of Equation (4.7) will allow us to obtain a second solution.

We ask: is there a function $w(x)$ such that

$$w(x) = x^\nu \sum_{n=0}^{\infty} b_n x^n \qquad(4.10)$$

with $\nu \neq 2$ and $b_0 \neq 0$?

We substitute this hypothetical solution into Equation (4.6) to obtain

$$(\nu^2 - \nu - 2)b_0 = 0, \qquad (\nu^2 + \nu - 2)b_1 = 0,$$

$$[(\nu + n)(\nu + n - 1) - 2]b_n = b_{n-2}, \qquad n > 1.$$

With $b_0 \neq 0$ and $\nu \neq 2$, it follows that

$$\nu = -1,$$

$$b_1 = 0,$$

$$b_2 = -b_0/2,$$

$$b_3 = \text{arbitrary (take as zero)}$$

$$\vdots$$

and, in general,

$$b_{2n} = \frac{b_{2n-2}}{2n(2n-3)} = \frac{-b_0(2n-1)}{(2n)!}. \qquad(4.11)$$

Hence, the solution

$$u_2(x) = \sum_{n=0}^{\infty} b_{2n} x^{2n-1} \qquad(4.12)$$

converges everywhere in $x \neq 0$. Once again, there are two linearly independent solutions of the second order differential equation, although one of them fails to exist at the singular point where the coefficient of the most highly differentiated term vanishes. In the above, we chose $b_3 = 0$. If this choice is not made, how does the added part of the solution compare with that previously obtained?

4.3 Problems

4.3.1 Let

$$u'' + \frac{1}{x}u' + \left(\alpha^2 - \frac{n^2}{x^2}\right)u = 0. \tag{4.13}$$

Find appropriate series expansions in powers of x for two linearly independent solutions of Equation (4.13) for each nonintegral real number n. Do we obtain two independent solutions when n is integral? We return to this matter in Problem 4.5.1.

4.3.2 Let

$$(1 - x^2)y''(x) - 2xy'(x) + \beta(\beta + 1)y(x) = 0. \tag{4.14}$$

Find appropriate series expansions in powers of x for two linearly independent solutions of Equation (4.14) for each real number β. Find such a pair of series expansions in powers of $(x - 1)$.

4.3.3 Let

$$xv'' + (1 - x)v' + nv = 0. \tag{4.15}$$

Find appropriate series expansions in powers of x for two linearly independent solutions of Equations (4.15) for each real number n.

4.3.4 Identify those values of β and of n in Problems 4.3.2 and 4.3.3 for which u and v are particularly interesting. Do the series obtained in Problems 4.3.1, 4.3.2, 4.3.3 converge for any values of x? What values?

4.3.5 Find series descriptions for the solutions of each of the following equations.

(a) $\qquad x^3 u'' + u = 0,$

(b) $\qquad u''' + xu = 0,$

(c) $\qquad x^4 u'' + (x - 1)u = 0,$

(d) $\qquad x^2 u'' - 3u = 0.$

For which of these do the series descriptions converge? In which are the coefficients of the series defined by two-term recurrence formulas (equations involving only 2 indices)? In which do the coefficients depend on higher order recurrence formulas? Can you conjecture a criterion which distinguishes those equations which have convergent series representations about the origin from those which may not? Can you identify a criterion to distinguish those equations whose series solutions will involve only two-term recurrence formulas?

4.4 Singular Points

The foregoing examples suggest rather strongly that, whenever the coefficients $\alpha(x)$, $\beta(x)$, of

$$u'' + \alpha(x)u' + \beta(x)u = 0 \qquad (4.16)$$

are such that

$$\alpha(x) = \sum_{n=0}^{\infty} \alpha_n(x - x_0)^n \qquad (4.17)$$

and

$$\beta(x) = \sum_{n=0}^{\infty} \beta_n(x - x_0)^n \qquad (4.18)$$

in some interval about x_0, two (and only two) linearly independent solutions of (4.16) exist and admit power series descriptions in some interval about x_0.* Conversely, when either or both of Equations (4.17), (4.18) are false, at least one of the linearly independent solutions seems not to be well behaved at x_0 and does not admit a power series representation about that point. We omit the proof† that each of these suggestions is correct, but go on to study a particular category of equations which arises naturally from the foregoing considerations.

We consider Equation (4.16) for any α, β, such that

$$x\alpha(x) = \sum_{n=0}^{\infty} \alpha_n x^n, \qquad (4.19)$$

$$x^2\beta(x) = \sum_{n=0}^{\infty} \beta_n x^n. \qquad (4.20)$$

As is suggested by the solutions of Equation (4.6), we ask whether, consistent with Equation (4.16), there are functions, $u(x)$, and numbers, ν, for which

$$u(x) = x^\nu \sum_{n=0}^{\infty} u_n x^n. \qquad (4.21)$$

---

* To put it more honestly, the reader can confirm that no contradictions to this suggestion occur when he tests it against all of the equations he has encountered.
† See George F. Carrier, Max Krook, Carl E. Pearson, *Functions of a Complex Variable, Theory and Technique*, McGraw-Hill, 1966 (reprinted by Hod Books, Ithaca, N.Y.), Chapter 5, for proofs of this and certain subsequent results.

The reader should verify that the problems described by Equations (4.16), (4.19), and (4.20) include those described by Equations (4.16), (4.17), and (4.18); moreover, equations in which $u''(x)$ has a variable coefficient [for example, Equation (4.6)] lie within the framework of this discussion.

When Equations (4.19), (4.20), and (4.21) are substituted into Equation (4.16), and the terms are regrouped appropriately, we obtain after division by $x^{\nu-2}$

$$[\nu(\nu - 1) + \alpha_0\nu + \beta_0]u_0 + \{(\nu + 1)\nu + \alpha_0(\nu + 1) + \beta_0\}u_1$$
$$+ (\alpha_1\nu + \beta_1)u_0\}x + \cdots + \{[(\nu + n)(\nu + n - 1)$$
$$+ \alpha_0(\nu + n) + \beta_0]u_n + \cdots\}x^n + \cdots = 0. \quad (4.22)$$

Therefore, in order that Equation (4.21) be a solution of Equation (4.16) and that $u_0 \neq 0$, it is certainly necessary that

$$\nu(\nu - 1) + \alpha_0\nu + \beta_0 = 0. \quad (4.23)$$

Equation (4.23) is termed the *indicial equation*. For each of the two values of ν implied by Equation (4.23), the values of the coefficients $u_1, u_2, \cdots, u_n, \cdots$ are implied by Equation (4.22); it would *seem* that the description of $u(x)$ as given by Equation (4.21) can be completed, *provided we take for granted the convergence of the right-hand side of Equation (4.21)*. We use the word *seem* advisedly since some difficulties can arise. We postpone a discussion of these difficulties and turn to the following examples.

4.5 Problems

4.5.1 Why need we not investigate the implications of Equation (4.22) when

$$u_0 = 0?$$

4.5.2 Will the analysis of this section lead to the same result if

$$x^\mu \alpha(x) = \sum_{n=0}^{\infty} \alpha_n x^n$$

with $\mu > 1$? With $\mu < 1$ and μ nonintegral?

Will the analysis of this section lead to the same result if

$$x^p \beta(x) = \sum_{n=0}^{\infty} \beta_n x^n?$$

with $p > 2$? With $p < 2$ and p nonintegral?

4.5.3 Try to find two series solutions of each of the equations

$$x^4 u''(x) + u(x) = 0, \qquad x^{1/2} v''(x) + v(x) = 0,$$

$$w''(x) + \frac{1}{x^2} w'(x) + \frac{1}{x} w(x) = 0.$$

Are there such series? If so, where do they converge?

4.5.4 Are there any values of α_0, β_0, for which Equations (4.21), (4.22), and (4.23) fail to provide two linearly independent solutions of Equation (4.16)?

[HINT: For what values of ν_1, ν_2 [the roots of Equation (4.23)] does anything drastic happen to the coefficients of the power series of one of the solutions? Why isn't Equation (4.6) such a special case?]

4.5.5 About what points, y, can a description of each solution of Equation (4.13) be found as a series of powers of $(x - y)$? Which choice of y would be most useful to

(1) describe each $u(x)$ in $x \geq 0$,
(2) describe each $u(x)$ in $0 \leq x \leq 2$,
(3) describe each $u(x)$ in $3 \leq x \leq 5$?

4.6 Singular Points—Continued

The differential equations we have just discussed, i.e., those satisfying conditions (4.19) and (4.20), with at least one of α_0, β_0, β_1 not zero, are those which are said to have a *regular singular point* at $x = 0$. When $\alpha(x)$ and $\beta(x)$ admit power series about $x = x_0$, x_0 is said to be an *ordinary point*. Points which are neither ordinary nor regular singular points are called (to no one's surprise) *irregular singular points*. The exercises of Problems 4.3 demonstrate that one cannot be assured of the existence of two solutions of a second order differential equation having the form (4.21) unless it (the differential equation) has either an ordinary point or a regular singular point at $x = x_0$. Even then, according to Problem 4.5.4, one solution may not be of this form.

Consider, in particular, the equation of Problem 4.3.1,

$$\frac{d^2 u(x, \beta)}{dx^2} + \frac{1}{x} \frac{du(x, \beta)}{dx} + \left(\alpha^2 - \frac{\beta^2}{x^2} \right) u(x, \beta) = 0. \qquad (4.24)$$

We found that, when $\beta \neq 0$,

$$u_1(x, \beta) = \sum_{m=0}^{\infty} \frac{(-1)^m (\alpha x/2)^{2m+\beta}}{m!(m+\beta)!/\beta!} \qquad (4.25)$$

and

$$u_2(x, \beta) = \sum_{m=0}^{\infty} \frac{(-1)^m (x/2)^{2m-\beta}}{m!(m-\beta)!/(-\beta)!} \tag{4.26}$$

are solutions of Equation (4.24), where, for present usage, it suffices to define $(m + a)!/a!$ as

$$(m + a)!/a! = (m + a)(m + a - 1)(m + a - 2) \cdots (a + 1). \tag{4.27}$$

We shall see later that this is consistent with the conventional definition of the factorial function $z!$.

When $\beta = 0$, these two solutions coincide, so that we need a new method for finding a second solution. One such method is to observe that, for $\beta \neq 0$,

$$U(x, \beta) = \frac{u_1(x, \beta) - u_2(x, \beta)}{\beta} \tag{4.28}$$

is a solution and it is possible that

$$Y(x) = \lim_{\beta \to 0} U(x, \beta) \tag{4.29}$$

is also a solution. This procedure motivates Problem 4.7.1.

4.7 Problems

4.7.1 Complete the process outlined above and show that $Y(x)$, as defined in Equation (4.29) is indeed a solution of Equation (4.24) with $\beta = 0$. Write Y in the form

$$Y = ?(x)u_1(x) + \sum_{n=0}^{\infty} v_n x^n,$$

where $?(x)$ is an elementary function you are to identify.

4.7.2 Carry out a similar process to obtain a second solution for the case $\beta = 1$.

The Wronskian | 5

The Wronskian, $W(x)$, of two functions, $u_1(x)$ and $u_2(x)$ is defined to be the value of the determinant

$$W(x) = \begin{vmatrix} u_1(x) & u_1'(x) \\ u_2(x) & u_2'(x) \end{vmatrix}. \tag{5.1}$$

The reader can verify that, if $u_1(x)$ and $u_2(x)$ are linearly dependent, on any interval $a < x < b$, the determinant vanishes for all x on that interval and the Wronskian is identically zero. Furthermore, if $W(x)$ vanishes for all x in an interval, and if one of the functions, say $u_1(x)$, is zero at no point on that interval, then

$$\frac{W}{u_1^2} = \frac{u_2'}{u_1} - u_1'\frac{u_2}{u_1^2} = 0,$$

so that u_2/u_1 is a constant, which implies linear dependence.

The Wronskian plays an important role in the study of second-order differential equations. Let [cf. Equations (3.1) and (3.2)]

$$[p(x)u'(x)]' + q(x)u(x) = 0, \tag{5.2}$$

and let $u_1(x)$ and $u_2(x)$ be two linearly independent solutions of Equation (5.2). It follows that we can write

$$u_2[(pu_1')' + qu_1] = 0 \tag{5.3}$$

and

$$u_1[(pu_2')' + qu_2] = 0. \tag{5.4}$$

Subtracting, we obtain

$$[p(u_1 u_2' - u_2 u_1')]' = 0$$

or

$$u_1 u_2' - u_2 u_1' = \text{const}/p(x). \tag{5.5}$$

Thus, only at points at which p is unbounded can the Wronskian of two linearly independent solutions of the equation vanish. One must bear in mind, of course, that at any singular point of Equation (5.2) either or both of u_1, u_2, may fail to exist. Thus, we generalize the implication of Equation (5.5) to state: *the Wronskian of two linearly independent solutions of a second-order linear homogeneous ordinary differential equation will vanish or fail to exist only at singular points of the differential equation.*

When one solution, $u_1(x)$ of Equation (5.1) is known, Equation (5.2) can be used to find a second linearly independent solution $u_2(x)$. Noting that

$$W = u_1^2(u_2/u_1)' = A/p(x), \tag{5.6}$$

we have

$$u_2(x) = A u_1(x) \int^x \frac{dx'}{p(x') u_1^2(x')}. \tag{5.7}$$

By choosing a particular nonzero value for A, and a particular lower limit of integration x_0, we get a specific member of the family of solutions given by Equation (5.7). One such choice is

$$u_2^*(x) = u_1(x) \int_{x_0}^x \frac{dx'}{p(x') u_1^2(x')},$$

where x_0 is some fixed number. Since the Wronskian of $u_1(x)$ and $u_2^*(x)$ is nonzero (what is it?), $u_1(x)$ and $u_2^*(x)$ are linearly independent. Thus, we have shown that if a second-order linear homogeneous equation, with $p(x)$ nonzero in an interval $a \leq x \leq b$, has one solution $u_1(x)$ which is nonzero in the interval, then there will be at least two linearly independent solutions in that interval. We have actually shown somewhat more, for since $u_2(x)$ in the derivation of Equation (5.7) was *any* solution of Equation (5.2), it follows that any solution of Equation (5.7) must be a linear combination of $u_1(x)$ and $u_2^*(x)$.

A consequence of this last statement is that our differential equation cannot have *more* than two linearly independent solutions. For suppose that $r(x)$, $s(x)$, and $t(x)$ are three solutions of Equation (5.2). Identify

$u_1(x)$ with $r(x)$, and construct $u_2^*(x)$ as above. Then, from the preceding paragraph, there must exist constants α, β, γ, δ, such that

$$s(x) = \alpha r(x) + \beta u_2^*(x), \qquad t(x) = \gamma r(x) + \delta u_2^*(x).$$

If β or δ is zero, we have already established linear dependence; if both β and δ are nonzero, then we multiply the first equation by δ and the second by β, and subtract, to demonstrate linear dependence between $r(x)$, $s(x)$, and $t(x)$.

Therefore, if a second-order linear homogeneous equation, with $p(x)$ nonzero in an interval $a \leq x \leq b$, has one nontrivial solution in that interval, then it has only two linearly independent solutions in that interval. Any pair of linearly independent solutions may be chosen as a *fundamental set* of solutions, and *any* other solution must then be expressible as a linear combination of the members of this set.

One might suppose that Equation (5.7) fails to provide information about $u_2(x)$ in the neighborhood of the zeros of $u_1(x)$. However, so long as u_1 vanishes at ordinary points of Equation (5.2), (i.e., where $p(x_j) \neq 0$) the limit of Equation (5.7) as $x \to x_j$ will give a meaningful result. In fact, the $u_2(x)$ so obtained will be continuous, differentiable, and very well behaved indeed at such points. This can be *expected* by noting that, when we investigated power series solutions of equations equivalent to Equation (5.2), we found no pathology in $u_2(x)$ at points where $u_1(x) = 0$. A few simple examples will strengthen the expectation that $u_2(x)$ does not misbehave at such points and careful analysis, which we omit (see, however, Exercise 5.1.6)) could be used to prove that no singular behavior of $u_2(x)$ arises except at singular points of Equation (5.2).

5.1 Problems

5.1.1 Let

$$(xu')' - (x - 3)u = 0, \qquad (5.8)$$

and verify that

$$u_1(x) = (1 - 2x)e^{-x}$$

is a solution of Equation (5.8). Use the technique of the foregoing Section to find a function $u_2(x)$ which is a solution of Equation (5.8) and is linearly independent of $u_1(x)$. Find $u(x)$ such that $u(0) = 1$; why do we need only one boundary condition?

5.1.2 Repeat (5.1.1) with
$$xy'' + 2y' + xy = 0 \qquad (5.9)$$
and
$$y_1(x) = x^{-1} \sin x.$$
Find the most general solution $y(x)$ for which $y(\pi/4) = 0$.

5.1.3 Show that there is a function $\sigma(x)$ such that, with the substitution
$$z = \int^x \sigma(x')\, dx',$$
Equation (5.2) can be recast in the form
$$h''(z) + F(z)h(z) = 0. \qquad (5.10)$$
Over what values of x can this be done?

5.1.4 Let $u_1(x)$ and $u_2(x)$ be two linearly independent solutions of Equation (5.2). Let $u_1(a) = u_1(b) = 0$, with $p(x) \neq 0$ and $u_1(x) \neq 0$ for $a < x < b$. By calculating $d(u_1/u_2)/dx$, show that $u_2(x)$ must have one zero in a, b; could it occur at one of the end points a, b? If one of a pair of linearly independent solutions is oscillatory over a certain region, what can you now state concerning the other one? What have you assumed concerning $p(x)$?

5.1.5 If $u_1(x)$ is one solution of Equation (5.2), define a new dependent variable $w(x)$ by means of $u(x) = u_1(x) \cdot w(x)$, and determine the equation satisfied by $w(x)$. Solve this equation, and compare your result with Equation (5.7).

5.1.6 Assume that $u_1(x_0) = 0$, and that $u_1(x) \cong B(x - x_0)$ for $|x - x_0|$ small. Let $p(x_0) \neq 0$. Evaluate the behavior of $u_2(x)$, as defined by Equation (5.7), for small values of $|x - x_0|$.

5.1.7 Construct a second-order linear homogeneous differential equation for which a fundamental set of solutions is given by the pair of functions $\sin x$ and $(1 - x)e^x$. Why does this result not violate the result of Problem 5.1.4?

Eigenvalue Problems | 6

We now extend our discussion of second-order homogeneous equations to cases in which a non-trivial solution satisfying certain boundary conditions exists only if a parameter of the equation is given a special value.

We study first the problem

$$(p(x)u')' + q(x)u + \lambda r(x)u = 0, \quad \text{in} \quad a < x < b \qquad (6.1)$$

with

$$u(a) = u(b) = 0; \qquad (6.2)$$

the functions p, q, and r are such that there are no singular points in $a \leq x \leq b$. We ask: are there any numbers, λ_n, and nontrivial functions, $u_n(x)$, such that when $\lambda = \lambda_n$, Equations (6.1) and (6.2) are satisfied by $u_n(x)$?

In answering this question, it is convenient to use the results of Problem 5.1.3 which allows us to rewrite Equation (6.1) in the form *

$$h''(z, \lambda) + Q(z)h(z, \lambda) + \lambda R(z)h(z, \lambda) = 0, \qquad (6.3)$$

where $h(z, \lambda) \equiv u(x)$, $z = z(x)$, $Q(z) = p(x)q(x)$, $R(z) = p(x)r(x)$, and primes denote differentiation with regard to z. We choose $z(a) = 0$ and we denote $z(b)$ by B.

---

*We adopt the notation $h(z, \lambda)$, instead of $h(z)$, both to emphasize the fact that h depends on λ and to simplify the phrasing of several important arguments which arise later in this section.

The results of the foregoing chapters establish that Equation (6.3) has two linearly independent solutions which we call $h_1(z, \lambda)$ and $h_2(z, \lambda)$, and we note that

$$H(z, \lambda) = \frac{h_1(0, \lambda)h_2(z, \lambda) - h_2(0, \lambda)h_1(z, \lambda)}{h_1(0, \lambda)h_2'(0, \lambda) - h_2(0, \lambda)h_1'(0, \lambda)} \qquad (6.4)$$

is that solution of Equation (6.3) for which $H(0, \lambda) = 0$ and $H'(0, \lambda) = 1$. The most general solution of Equation (6.3) for which $h(0, \lambda) = 0$ is a constant multiple of $H(z, \lambda)$.

For any function $Q(z)$ which is bounded in $0 \leq z \leq B$ and any function $R(z)$ which is positive * in $0 \leq z \leq B$, there is a number, A, which is large enough that, when $\lambda \geq A$, $Q(z) + \lambda R(z) > 0$ for each z in

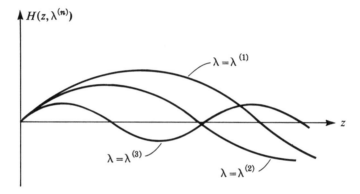

Figure 6.1 $H(z, \lambda^{(n)})$ versus z for three values of n.

$0 \leq z \leq B$. For three such values of λ, with $\lambda^{(3)} > \lambda^{(2)} > \lambda^{(1)}$, the functions $H(z, \lambda^{(n)})$ are sketched in Figure 6.1. According to Equation (6.3), the curvature of each curve is positive when H is negative and negative when H is positive. Accordingly, we expect each function, $H(z, \lambda^{(n)})$ to be oscillatory and to vanish at several values of z. Furthermore, our experience with the constant-coefficient equation and with Problem 3.3.7 suggests that the undulations are more frequent for larger average values of $Q(z) + \lambda R(z)$ (cf. also Problem 6.1.8). Thus, we expect with some confidence that the roots of $H(z, \lambda^{(3)})$ are closer together than those of $H(z, \lambda^{(2)})$, etc., as shown in the sketch. We also expect that the position of each zero of $H(z, \lambda)$ moves continuously to

---

* The reader should reword what follows to cover the situation in which $R(z) < 0$ in $0 < z < B$.

the left as λ increases. That is, if the location of the nth root of $H(z, \lambda) = 0$ is called $z_n(\lambda)$, we expect each function $z_n(\lambda)$ to decrease continuously and monotonically as λ increases. This implies that there are many values (how many?) of λ for which $H(B, \lambda) = 0$. It also implies that when we call these values λ_n and order the λ_n so that $\lambda_{n+1} > \lambda_n$, each of the corresponding functions $H(z, \lambda_n)$ will have one more zero in the interval $0 < z < B$ than its predecessor. This ordered set of numbers, λ_n, is the set of eigenvalues of the problem defined by Equation (6.1) and (6.2), and the functions $u_n(x)$, defined by

$$u_n(x) = H(z, \lambda_n),$$

is the set of eigenfunctions.

It is important to realize that, when any point in $a \leq z \leq b$ is a singular point of Equation (6.1), when the interval $a < x < b$ is not finite, or when $r(x)$ is not one-signed in $a < x < b$, the situation *may* be much more complicated and the reader is referred to the more sophisticated treatment by C. E. Titchmarsh, *Eigenfunctions and Eigenfunction Expansions*, Oxford, Clarendon Press, 1946. The fact that *some* singular problems are *not* more complicated can be concluded by studying Problem 6.1.5.

Suppose, now, that the foregoing expectations are realized in some particular problem of the form given by Equation (6.1) and (6.2). We denote the eigenvalues of that problem by λ_n and the eigenfunctions by $\phi_n(x)$. It is profitable to consider the following manipulation. We write

$$\phi_m(x)[(p\phi_n')' + q\phi_n + \lambda_n r\phi_n] = 0, \tag{6.5}$$

$$\phi_n(x)[(p\phi_m')' + q\phi_m + \lambda_m r\phi_m] = 0, \tag{6.6}$$

where the vanishing of the brackets is required by the definition of λ_j and ϕ_j. We subtract Equations (6.5) and (6.6) and identify the results as

$$[p(\phi_m \phi_n' - \phi_n \phi_m')]' + (\lambda_n - \lambda_m)r\phi_n \phi_m = 0. \tag{6.7}$$

We integrate Equation (6.7) over the interval $a < x < b$ to obtain

$$(\lambda_m - \lambda_n) \int_a^b r(x)\phi_n(x)\phi_m(x)\, dx = 0$$

since each of ϕ_m, ϕ_n vanish at a and b. Thus, for any m and any $n \neq m$,

$$\int_a^b r(x)\phi_n(x)\phi_m(x)\, dx = 0, \tag{6.8}$$

and the eigenfunctions ϕ_n and ϕ_m are said to be orthogonal. Note that the meaning of the word orthogonal depends on the coefficient, $r(x)$;

$r(x)$ is commonly referred to as the *weighting function* in connection with its use in Equation (6.8).

It frequently is convenient to choose the arbitrary multiplication constants occurring in the eigenfunctions $\phi_n(x)$ so that

$$\int_a^b r(x)\phi_n^2 \, dx = 1.$$

When this choice is made, the eigenfunctions are said to have been *normalized*. The resulting set of normalized functions $\phi_n(x)$ is said to be *orthonormal*.

The idea of orthogonality encountered in Equation (6.8) is a generalization of the conventional definition used for two vectors in three-dimensional space. If A and B are two such vectors, with components (A_1, A_2, A_3) and (B_1, B_2, B_3) respectively, then it is a familiar fact that these two vectors are orthogonal if their scalar product vanishes, i.e., if $A_1 B_1 + A_2 B_2 + A_3 B_3 = 0$. Similarly, two vectors in n-dimensional space, with components $(A_1, A_2, \ldots, A_n)$ and $(B_1, B_2, \ldots, B_n)$ are said to be orthogonal if the generalized scalar product $A_1 B_1 + A_2 B_2 + \cdots + A_n B_n = 0$. If, still more generally, we are given two functions $f(x)$ and $g(x)$ defined over an interval (a, b) and we use the usual interpretation

$$\int_a^b f(x)g(x) \, dx = \lim_{n \to \infty} \sum_{i=1}^n f(x_i) \cdot g(x_i) \, \Delta x,$$

it becomes natural to think of the integral on the left-hand side as being a sort of continuous scalar product of the two functions $f(x)$ and $g(x)$; the functions would be said to be orthogonal if $\int_a^b f(x)g(x) \, dx = 0$. Our final generalization consists in the incorporation of a weighting function, $r(x)$ [which we take as >0 in (a, b)], and this leads to the use of the word *orthogonality* in connection with Equation (6.8).

In the same way, the *length* or *norm*, $\|f\|$, of a function $f(x)$ is defined as the square root of the generalized scalar product of the function (thought of as a vector in *function space*) with itself; thus

$$\|f\|^2 = \int_a^b r(x) f^2(x) \, dx. \tag{6.9}$$

6.1 Problems

6.1.1 Where, in the above Section, did we need the statement, "p, q, and r are such that there are no singular points in $a \leq x \leq b$"? Where do we need the requirement $r(x) > 0$ in $a \leq x \leq b$?

6.1.2 Modify the manipulations of this section on eigenvalue problems so that the results apply for any boundary conditions of the form

$$\alpha_1 u'(a) + \alpha_2 u(a) = 0, \qquad \beta_1 u'(b) + \beta_2 u(b) = 0, \tag{6.10}$$

where at least one of α_1, α_2 and at least one of β_1, β_2 are not zero. This general problem is called the *Sturm-Liouville* problem. Consider also the case of *periodic* end conditions: $p(a) = p(b)$, $u(a) = u(b)$, and $u'(a) = u'(b)$.

6.1.3 Find the eigenvalues and eigenfunctions of

$$(x^2 \phi')' + \lambda x^2 \phi = 0 \tag{6.11}$$

with

$$\phi'(0) = \phi(1) = 0.$$

6.1.4 Find the eigenvalues and eigenfunctions of Equation (6.11) with $\phi(1) = \phi(3) = 0$. Repeat with $\phi'(0) = 0$, $\phi(1) + \phi'(1) = 0$.

6.1.5 Why do Problems 3.3.6 and 6.1.3 lie outside the scope of the development of this section? Are their eigenfunctions orthogonal?

6.1.6 Replace ϕ_m in Equation (6.5) by ϕ_n and deduce that

$$\lambda_n = \frac{\int_a^b [p(\phi_n')^2 - q\phi_n^2]\,dx}{\int_a^b r\phi_n^2\,dx}.$$

Are there situations in which you could now show that $\lambda_n > 0$?

6.1.7 Are there any advantages in framing an eigenvalue discussion in terms of Equation (6.1), rather than in terms of

$$a_2(x)y'' + a_1(x)y' + [a_0(x) + \lambda b_0(x)]y = 0?$$

If such an equation is transformed into the form of Equation (6.1), how are the new and old eigenfunctions and eigenvalues related to one another?

6.1.8 Let y_1 and y_2 satisfy the differential equations

$$y_1'' + A_1(x)y_1 = 0, \qquad y_2'' + A_2(x)y_2 = 0,$$

with $A_2(x) > A_1(x)$ in (α, β). Let $y_1(\alpha) = y_1(\beta) = 0$, with $y_1(x) \neq 0$ for x in (α, β). Let $y_2(\alpha) = 0$; show that $y_2(x)$ must vanish for some value of x satisfying $\alpha < x < \beta$. [HINT: Multiply the first equation by y_2, the second by y_1, subtract, and integrate from α to β.]

6.2 Fourier Series

We turn now to one of the central topics in Sturm-Liouville theory, the possibility of expanding some given function $\psi(x)$, over an interval (a, b), in a series of normalized eigenfunctions $\phi_n(x)$ of some Sturm-Liouville problem involving that interval. If such an expansion

$$\psi(x) = \sum_{n=0}^{\infty} k_n \phi_n(x) \tag{6.12}$$

were valid, it would represent a generalization of the familiar power series expansion

$$u(x) = \sum_{n=0}^{\infty} a_n x^n,$$

feasible for many functions (including, as we have seen, the solutions of certain differential equations). In terms of our function-space interpretation, the right-hand side of Equation (6.12) would represent a decomposition of $\psi(x)$ into components along the mutually perpendicular unit vectors $\phi_n(x)$.

If Equation (6.12) is valid, then we can determine the k_n by multiplying both sides of the equation by $r(x)\phi_m(x)$ (for some choice of m) and integrating over (a, b). Because of the orthonormal character of the $\phi_n(x)$, this process gives

$$\int_a^b r(x)\psi(x)\phi_m(x)\,dx = k_m. \tag{6.13}$$

However, there is another, rather instructive, approach to this formula. Suppose that we try to approximate $\psi(x)$, over the interval (a, b), by a linear combination of the first N eigenfunctions $\phi_n(x)$. How do we choose the k_n so that, in the statement

$$\psi(x) \cong k_1\phi_1(x) + k_2\phi_2(x) + \cdots + k_N\phi_N(x)$$

the discrepancy function

$$f(x) = \psi(x) - [k_1\phi_1(x) + \cdots + k_N\phi_N(x)]$$

is as small as possible over the interval (a, b)? Since we have defined the size of a function $f(x)$ by Equation (6.11), it is natural to try to choose the k_n so as to make

$$I(k_1, k_2, \ldots, k_n) = \int_a^b r(x)[\psi(x) - \{k_1\phi_1(x) + \cdots + k_N\phi_N(x)\}]^2\,dx \tag{6.14}$$

as small as possible. (Since we still require $r(x) > 0$ in (a, b), we note parenthetically that I can be small only if the approximation $\psi(x) \cong k_1\phi_1 + \cdots + k_n\phi_n$ is a good one.) Defining

$$\int_a^b r(x)\psi(x)\phi_n(x) = c_n,$$

we obtain

$$I(k_1, k_2, \ldots, k_n)$$
$$= \int_a^b r(x)\psi^2(x)\,dx - 2\int_a^b r(x)\psi(x)[k_1\phi_1(x) + \cdots + k_N\phi_N(x)]\,dx$$
$$+ \int_a^b r(x)[k_1\phi_1(x) + \cdots + k_N\phi_N(x)]^2\,dx$$
$$= \int_a^b r(x)\psi^2(x)\,dx - 2(k_1c_1 + \cdots + k_N c_N) + (k_1^2 + \cdots + k_N^2)$$
$$= \int_a^b r(x)\psi^2(x)\,dx + k_1(k_1 - 2c_1) + \cdots + k_N(k_N - 2c_N). \quad (6.15)$$

Now as the k_n are varied, each expression of the form $k_n(k_n - 2c_n)$ attains its minimum when $k_n = c_n$; thus, the least value of $I(k_1, \ldots, k_n)$ is attained when the k_n satisfy Equation (6.13).

We conclude that whenever we wish to approximate a function $\psi(x)$ in the sense implied by Equation (6.14), the prescription for finding the coefficients is just Equation (6.13) and does not depend on the number of terms in the description. *Although it is beyond the scope of this work to show it, it is true that, for eigenfunctions $\phi_n(x)$ of a nonsingular Sturm-Liouville problem*, we have

$$\lim_{N \to \infty} \int_a^b r(x)\left(\psi(x) - \sum_{n=0}^N k_n\phi_n\right)^2 dx = 0, \quad (6.16)$$

provided that the k_n are chosen by the rule (6.13) and that $\int_a^b \psi^2(x)\,dx$ exists. (Note that $\psi(x)$ need not satisfy the boundary conditions satisfied by the individual eigenfunctions $\phi_n(x)$.)

It is in this sense (and this sense only) that, when we write

$$\psi(x) = \sum_{n=0}^\infty k_n\phi_n(x),$$

we can assert that the series converges to $\psi(x)$. Conventional language for this is: the series converges to $\psi(x)$ in the mean square sense.

It is also true that, at points x_ν where $\psi(x)$ is not continuous, $\sum_{n=0}^N k_n\phi_n(x_\nu)$ approaches the value

$$\tfrac{1}{2}[\psi(x_\nu-) + \psi(x_\nu+)] \quad \text{as } N \to \infty, \quad (6.17)$$

where $\psi(x_\nu+)$ and $\psi(x_\nu-)$ denote the limits of $\psi(x)$ as $x \to x_\nu$ from the right and left respectively.

6.3 Problems

6.3.1 Find the eigenfunction expansion of
$$\psi(x) = 1 - e^{-\alpha x}$$
in
$$0 < x < \pi$$
in terms of the eigenfunctions of
$$\phi'' + \lambda\phi = 0 \quad \text{in} \quad 0 < x < \pi$$
with
$$\phi(0) = \phi'(\pi) = 0.$$

For what values of α is the eigenfunction expansion an effective description of $\psi(x)$?

6.3.2 Find the eigenfunction expansions of

(a) $\qquad f_1(x) = \sin^2 x,$
(b) $\qquad h_1(x) = 1,$
(c) $\qquad h_2(x) = \pi^2 - x^2,$
(d) $\qquad h_3(x) = (\pi^2 - x^2)^2,$

in the interval $0 < x < \pi$.

Use the eigenfunctions of $u'' + \lambda u = 0$ with $u(0) = u(\pi) = 0$. Study the results and infer what you can.

6.3.3 Call the eigenfunctions of a given Sturm-Liouville system $\chi_j(x)$. Let the interval be $0 < x < 1$ and the weighting function be $r(x)$ as in Equation (6.1). Furthermore, let

$$\psi(x, \epsilon) = \begin{cases} \dfrac{1}{2\epsilon} & \text{for } 0 < a - \epsilon < x < a + \epsilon < 1, \\ 0 & \text{for all other } x. \end{cases}$$

Find the expansion of $\psi(x, \epsilon)$ in the $\chi_j(x)$ and, in particular, find the limit obtained when, in the series for ψ, $\epsilon \to 0$.

6.3.4 (a) Show that all eigenvalues and eigenfunctions of Equation (6.1), with $r(x) > 0$ in (a, b), must be real.

(b) With the k_n defined as in Equation (6.13), show that $\sum_{j=1}^{\infty} k_j^2$ converges.

6.4 A Special Example

We now terminate our eigenvalue discussion, for the present, by considering a rather unusual but illustrative example.

Let
$$u''(x) + \lambda F(x)u(x) = 0 \quad \text{in} \quad -1 < x < 1 \quad (6.18)$$

with
$$u(-1) = u(1) = 0$$

and with
$$F(x) = \begin{cases} \dfrac{1}{2\epsilon}, & |x| < \epsilon \ll 1, \\ 0, & |x| > \epsilon. \end{cases}$$

It is clear that
$$u(x) = A(1 + x) \quad \text{in} \quad -1 < x < -\epsilon$$

and $\hspace{7cm}$ (6.19)
$$u(x) = B(1 - x) \quad \text{in} \quad \epsilon < x < 1.$$

Furthermore,
$$u = C \cos(x\sqrt{\lambda/2\epsilon}) + D \sin(x\sqrt{\lambda/2\epsilon}) \quad \text{in} \quad |x| < \epsilon. \quad (6.20)$$

In view of the symmetry of the problem, we can expect odd eigenfunctions, or even eigenfunctions, but *no* eigenfunctions which are neither even nor odd in x (cf. Problem 6.5.3).

We look first for even eigenfunctions. For these, $A = B$ and $D = 0$. The requirement that u and u' be continuous at $x = \epsilon$ leads to

$$A(1 - \epsilon) = C \cos \sqrt{\lambda\epsilon/2}$$

and $\hspace{7cm}$ (6.21)

$$-A = -C\sqrt{\lambda/2\epsilon} \sin \sqrt{\lambda\epsilon/2},$$

so that
$$1 - \epsilon = \sqrt{2\epsilon/\lambda} \cot \sqrt{\lambda\epsilon/2}.$$

Call

$$\sqrt{\lambda\epsilon/2} = z.$$

Then

$$\cot z = \frac{1-\epsilon}{\epsilon} z. \qquad (6.22)$$

Figure 6.2 displays clearly that there is one root,* z_0, for which $z_0 \ll 1$ and that all other roots, z_n, lie close to $z_n = n\pi$.

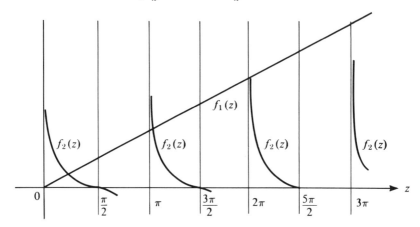

Figure 6.2 Graphs of $f_1(z) = (1 - \epsilon)z/\epsilon$ and $f_2(z) = \cot z$ to illustrate location of roots of $f_1(z) = f_2(z)$.

Since $z_0 \ll 1$, then $\cot z_0 \cong 1/z_0$, and Equation (6.2) reduces to

$$z_0^2(1-\epsilon) \cong \epsilon \quad\text{and}\quad z_0 \cong \epsilon^{1/2},$$

so that

$$\lambda_0 \cong 2.$$

For the other roots, we write $z_n = n\pi + \beta_n$ so that [anticipating, via Figure 6.2, that $\beta_n \ll 1$], Equation (6.22) becomes

$$(n\pi + \beta_n)(1-\epsilon) \cong \epsilon/\beta_n \quad\text{and}\quad \beta_n \cong \epsilon/n\pi.$$

Thus, the eigenvalues are approximately 2, $2\pi^2/\epsilon$, $8\pi^2/\epsilon$, $18\pi^2/\epsilon$, ...

The reader can substitute these results into the description of $u(x)$

---

* We need consider only the positive roots, since the negative roots are the same, except for sign, and to obtain λ we must square z.

(i.e., into Equations (6.19) and (6.20), using Equation (6.21)) and verify that the sketches of Figure 6.3 (for the choice $A = 1$) characterize the eigenfunctions of the problem.

It is especially interesting to ask what happens as $\epsilon \to 0$. We see from the above that

$$\lambda_0 \to 2, \quad \lambda_n \to \infty \quad \text{for } n \neq 0,$$

and

$$u_n(x) \to \begin{cases} 1 + x, & x < 0, \\ 1 - x, & x > 0 \end{cases}$$

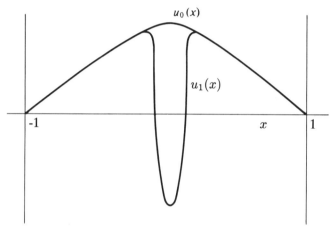

Figure 6.3 Sketches of first two even eigenfunctions.

as $\epsilon \to 0$. (Remember that u_n differs from a pair of straight lines only for $|x| < \epsilon$.) However, for any $\epsilon > 0$, $u_n(x)$ is still oscillatory for $|x| < \epsilon$, so that $u_n'(x)$ has a nonuniform limit as $\epsilon \to 0$.

6.5 Problems

6.5.1 Without solving the differential equation, characterize and sketch your expectation of the appearance of several of the eigenfunctions of

$$u'' + \lambda f u = 0 \quad \text{with} \quad u(0) = u(3) = 0.$$

The function f is defined by

$$f(x) = \begin{cases} 1, & 0 < x < 1, \\ 0, & 1 < x < 2, \\ 1, & 2 < x < 3. \end{cases}$$

6.5.2 Find the eigenfunctions and eigenvalues of
$$x^2 u'' + \lambda u = 0 \quad \text{in} \quad 1 < x < 2 \quad \text{with} \quad u(1) = u'(2) = 0.$$

6.5.3 Verify that there are no eigenfunctions for the problem accompanying Equation (6.18) except the even and the odd ones. Find the odd eigenfunctions of that problem.

6.5.4 Try to *approximate* some of the eigenvalues and eigenfunctions of
$$u'' + \lambda e^{-Ax^2} u = 0 \quad \text{with} \quad u(-1) = u(1) = 0,$$
where $A \gg 1$.

How many of the $\phi_n(x)$ you find are likely to be good approximations? Use any definition of "good" you like, but state it.

6.5.5 Find the eigenfunction expansions of $f(x) = 2 - x^2$ and of $g(x) = x$, using the eigenfunctions of the problem associated with Equation (6.18). Why does the sum of the first few terms of each series differ so much from $f(x)$ and $g(x)$?

6.5.6 The hairspring of a particular watch is so designed that the angular acceleration of the balance wheel is $-k$ times its angular displacement from a given reference position. Furthermore, each time the balance wheel arrives at this reference position, the mainspring immediately imparts an angular velocity, Ω, to the balance wheel. Assuming that your watch is of this design, listen to it and find values of k and Ω which are consistent with the foregoing description and with your observation. What observational fact do you need?

The manufacturing tolerances are such that k can differ from its design value by αk and Ω can differ from its design value by $\beta\Omega$, where $0 < \alpha \ll 1$ and $0 < \beta \ll 1$. What kind of adjustment would you suggest? How should the adjustment range depend on α and β?

The Second-Order Linear Nonhomogeneous Equation | 7

The method which led to Equation (5.7) can be extended to give a systematic and useful description of the solution of the nonhomogeneous equation; we turn now to that extension. Of the various forms into which the general second order linear equation can be put (cf Equations (3.1) through (3.5)) that corresponding to Equation (3.2) leads to the simplest results. Thus, let

$$[p(x)\phi'(x)]' + q(x)\phi(x) = g(x) \quad \text{in} \quad a < x < b, \tag{7.1}$$

and let $\psi_1(x)$ and $\psi_2(x)$ be linearly independent solutions of the homogeneous equation. As usual, we suppose that $p(x)$ does not vanish in (a, b). We define

$$W_1(x) = \psi_1(x)\phi'(x) - \phi(x)\psi_1'(x)$$

and use again the now-familiar device wherein we write

$$\psi_1\{(p\phi')' + q\phi\} = \psi_1 g, \tag{7.2}$$

$$\phi\{(p\psi_1')' + q\psi_1\} = 0. \tag{7.3}$$

We subtract Equation (7.3) from Equation (7.2) to obtain

$$(pW_1)' = \psi_1 g.$$

It follows that

$$p(x)W_1(x) - p(x_0)W_1(x_0) = \int_{x_0}^{x} \psi_1(t)g(t)\,dt, \tag{7.4}$$

59

where x_0 is any point such that $a < x_0 < b$. Defining
$$W_2(x) = \psi_2(x)\phi'(x) - \psi_2'(x)\phi(x)$$
and proceeding as above, we also obtain
$$p(x)W_2(x) - p(x_0)W_2(x_0) = \int_{x_0}^{x} \psi_2(t)g(t)\,dt. \tag{7.5}$$

We subject Equations (7.4) and (7.5) to our "multiply and subtract" device again, and write
$$\psi_2(x)[p(x)W_1(x) - p(x_0)W_1(x_0)] - \psi_1(x)[p(x)W_2(x) - p(x_0)W_2(x_0)]$$
$$= \int_{x_0}^{x} [\psi_2(x)\psi_1(t) - \psi_1(x)\psi_2(t)]g(t)\,dt.$$

This last equation may be re-written as
$$p(x)W(x)\phi(x)$$
$$= C_1\psi_1(x) + C_2\psi_2(x) + \int_{x_0}^{x} [\psi_2(x)\psi_1(t) - \psi_1(x)\psi_2(t)]g(t)\,dt, \tag{7.6}$$
where
$$C_1 = -p(x_0)W_2(x_0), \quad C_2 = p(x_0)W_1(x_0) \quad \text{and} \quad W(x) = \psi_1\psi_2' - \psi_2\psi_1'.$$

We have already learned in Chapter 5 that $W(x)p(x)$ is a constant whose value is implied as soon as we choose the linearly independent solutions ψ_1, ψ_2 of the homogeneous equation. If we denote this constant by C and recall that all solutions of Equation (7.1) can be written as the sum of a particular solution of Equation (7.1) and any linear combination of its homogeneous solutions, we see that all solutions of Equation (7.1) have the form
$$\phi(x) = A_1\psi_1(x) + A_2\psi_2(x) + \frac{1}{C}\int_{x_0}^{x} [\psi_2(x)\psi_1(t) - \psi_1(x)\psi_2(t)]g(t)\,dt. \tag{7.7}$$

As an example of the foregoing we examine the problem
$$w''(y) - (1 + y^2)w(y) = y, \tag{7.8}$$
and we seek that solution for which $w(0) = a$, $w'(0) = b$. The reader may verify that two linearly independent solutions of the homogeneous equation are
$$\psi_1(y) = e^{y^2/2}, \quad \psi_2(y) = e^{y^2/2}\operatorname{erf} y,$$
where the function $\operatorname{erf} y$ (the error function) is defined by
$$\operatorname{erf} y = \frac{2}{\sqrt{\pi}}\int_0^{y} e^{-\alpha^2}\,d\alpha. \tag{7.9}$$

The reader will recall that he encountered this function (and graphed it) in Problem 1.6.2. He should also note that erf $(\infty) = 1$.

The Wronskian of ψ_1 and ψ_2 is

$$W = \psi_1 \psi_2' - \psi_2 \psi_1' = 2/\sqrt{\pi}$$

and, since in this problem the function corresponding to $p(x)$ in Equation (7.6) is $p(y) = 1$, the application of Equation (7.7) to this problem gives

$$w(y) = A_1 \psi_1(y) + A_2 \psi_2(y) + \frac{\sqrt{\pi}}{2} \int_0^y \exp\left(\frac{y^2 + t^2}{2}\right)(\text{erf } y - \text{erf } t) t \, dt,$$

where A_1 and A_2 are constants to be determined by the boundary conditions. Integration by parts yields

$$w(y) = A_1 e^{y^2/2} + A_2 e^{y^2/2} \text{ erf } y + \frac{\sqrt{\pi}}{2} e^{y^2/2} \left[\sqrt{2} \text{ erf } \frac{y}{\sqrt{2}} - \text{erf } y\right].$$

The boundary conditions require that

$$w(0) = A_1 = a, \quad w'(0) = 2A_2/\sqrt{\pi} = b,$$

so that $w(y)$ is uniquely determined.

7.1 Problems

7.1.1 Use the foregoing technique to show that

$$\chi(x) = \int_0^x \frac{\sin[\lambda(x - t)]}{\lambda} h(t) \, dt$$

is a solution of

$$\chi''(x) + \lambda^2 \chi(x) = h(x).$$

(a) For $h(x) = x$, $\chi(1) = 0$, and $\chi'(1) = 1$, find $\chi(x)$ in $1 < x$.

(b) Let $h(x) = \sin x$, $\chi(0) = 0$ and $\chi(\pi) = 0$; find $\chi(x)$ for *every* real value of the parameter λ.

7.1.2 Let

$$(xu')' + (3 - x)u = (1 - 2x)e^{-x}$$

and note that

$$f_1(x) = (1 - 2x)e^{-x}$$

is a solution of the homogeneous equation. Using the Wronskian method and some integrations by parts, show that a second solution of the homogeneous equation is (as you found in Exercise 5.1.1)

$$f_2(x) = e^x + (1 - 2x)e^{-x}E(2x),$$

where

$$E(2x) = \int_1^{2x} \frac{e^u}{u}\, du.$$

Continue the technique and find all $u(x)$ (if there are any) such that

(a) $u(0) = 1,$

(b) $u(0) = 0,$

(c) $u(\infty) = 0,$

(d) $u(0) = 1$ and $u(\infty) = 0,$

(e) $u(0) = 1$ and $u(1) = 0,$

(f) $u(\tfrac{1}{2}) = \sqrt{e}.$

7.1.3 Let

$$x[xu'(x, t, \epsilon)]' - u(x, t, \epsilon) = f(x, t, \epsilon) \quad \text{in} \quad 1 < x < 2$$

with

$$f(x, t, \epsilon) = \begin{cases} \dfrac{1}{2\epsilon}, & 1 < t - \epsilon < x < t + \epsilon < 2, \\ 0 & \text{for all other } x, \end{cases}$$

and with $u(1, t, \epsilon) = u'(2, t, \epsilon) = 0$. Primes denote differentiation with respect to x. Find $u(x, t, \epsilon)$ and, in particular, find

$$\lim_{\epsilon \to 0} u(x, t, \epsilon) = U(x, t).$$

Find

$$U'(t+, t) - U'(t-, t).$$

7.1.4 Let the interval $1 < x < 2$ be divided into intervals of length $2\epsilon = 1/M$, where M is some integer. Let $t_1, t_2, \ldots, t_M$ be the midpoints of these intervals, and let

$$x(xu')' - u = 2\epsilon \sum_{n=1}^{M} a_n f(x, t_n, \epsilon) \quad \text{in} \quad 1 < x < 2,$$

with $u(1, \epsilon) = u'(2, \epsilon) = 0$. (Why don't we include t_n as an argument of u?) Here f is the function of Problem 7.1.3, and the a_n are constants. Write $u(x, \epsilon)$ as a sum, using the results of Problem 7.1.3, and find the limit as $M \to \infty$. In this calculation, let a_n be the value of the continuous function $a(x)$ evaluated at $x = t_n$, that is, $a_n = a(t_n)$. Your answer should be the solution, for some function $g(x)$, of the boundary-value problem

$$x(xu')' - u = g(x) \quad \text{in } 1 < x < 2$$

with $u(1) = u'(2) = 0$. What is that function, $g(x)$?

7.1.5 Find and sketch a graph of that solution of Equation (7.8) for which $w(0) = 0$ and $w(y)$ is bounded as $y \to \infty$. Use a table of transcendental functions * for the evaluation of the function erf y.

7.1.6 Two alternative approaches to the solution of the nonhomogeneous equation

$$a_2(x)y'' + a_1(x)y' + a_0(x)y = g(x)$$

are outlined in parts (a) and (b). Show that each approach does lead to the desired result, and make appropriate comparisons with Equation (7.7).

(a) Let $u(x)$, $v(x)$ be a pair of linearly independent solutions of the homogeneous counterpart. Write $y = \alpha(x)u(x) + \beta(x)v(x)$, where $\alpha'u + \beta'v \equiv 0$, and substitute for y in the differential equation. (This is called the *method of variation of parameters*.)

(b) Let $u(x)$ be one solution of the homogeneous counterpart. Write $y = u(x)\,w(x)$, and substitute for y in the differential equation.

7.1.7 Let the eigenvalues and eigenfunctions of the problem

$$(p\phi')' + (q + \lambda r)\phi = 0, \qquad \phi(a) = \phi(b) = 0$$

be denoted by $\lambda_1, \lambda_2, \lambda_3, \ldots$ and $\phi_1, \phi_2, \phi_3, \ldots$, respectively. Show that if λ has one of the values λ_n, then the problem

$$(p\phi')' + (q + \lambda_n r)\phi = g(x), \qquad \phi(a) = \phi(b) = 0$$

cannot have a solution if $\int_a^b \phi_n g \, dx \neq 0$. Extend this result so as to apply to boundary conditions of the more general Sturm-Liouville type.

---

* See for example Jahnke and Emde, *Tables of Functions*, Dover, 1945 or *Handbook of Mathematical Functions*, National Bureau of Standards Appl. Math. Series 55, June, 1964 (Wiley reprint, 1972).

7.2 Green's Functions

Problems 7.1.3 and 7.1.4 suggest rather strongly that the solution in an interval $a < x < b$ of a second-order nonhomogeneous equation

$$[p(u')]' + qu = g(x) \qquad (7.10)$$

with homogeneous boundary conditions at $x = a$ and $x = b$ can be constructed in the form

$$u = \int_a^b K(x, t) g(t)\, dt. \qquad (7.11)$$

They also suggest that $K(x,t)$ is a function which, in $a < x < t$, is a solution $V_1(x)$ of the homogeneous counterpart of (7.10); that, in $t < x < b$, is another solution $V_2(x)$ of that homogeneous equation; that $V_1(t) = V_2(t)$; that $V_1(a)$ obeys the boundary condition on $u(a)$; that $V_2(b)$ obeys the boundary condition on $u(b)$; and that $V_1'(t) \neq V_2'(t)$. To test this suggestion and to find a simple way to construct $K(x,t)$, we study the specific problem in which we seek a solution of Equation (7.10) with $u(a) = u(b) = 0$. The reader will see that the procedure differs only superficially for any alternative homogeneous boundary conditions (one at a and one at b). We recall that if $V_1(x)$ is a solution of $(pu')' + qu = 0$ and $V_1(a) = 0$, the *only* solutions of $(pu')' + qu = 0$ with $u(a) = 0$ are constant multiples of $V_1(x)$. Thus

$$K(x, t) = A V_1(x) \qquad \text{in} \quad a < x < t.$$

Furthermore, if $V_2(x)$ is a solution for which $V_2(b) = 0$, $K(x,t)$ must be

$$K(x, t) = B V_2(x) \qquad \text{in} \quad t < x < b.$$

$K(x, t)$ will be continuous at $x = t$ only if

$$A V_1(t) = B V_2(t),$$

and we conclude that *

$$K(x, t) = \begin{cases} C V_2(t) V_1(x) & \text{in} \quad a < x < t, \\ C V_2(x) V_1(t) & \text{in} \quad t < x < b. \end{cases} \qquad (7.12)$$

With this information at our disposal we ask: is there a number C such that

---

* Note that A and B are constants only in the usual sense that they do not vary with x.

$$u(x) = \int_a^b K(x, t)g(t)\, dt = \int_a^x CV_1(t)V_2(x)g(t)\, dt + \int_x^b CV_1(x)V_2(t)g(t)\, dt \tag{7.13}$$

is the solution to the stated boundary-value problem? To answer the question we differentiate Equation (7.13) with care to obtain

$$u'(x) = \int_a^x CV_2'(x)V_1(t)g(t)\, dt + \int_x^b CV_2(t)V_1'(x)g(t)\, dt$$

and

$$[pu'(x)]' = Cg(x)p(x)W(x) + C[p(x)V_2'(x)]' \int_a^x V_1(t)g(t)\, dt$$
$$+ C[p(x)V_1'(x)]' \int_x^b V_2(t)g(t)\, dt,$$

where the Wronskian, W, is given by

$$W = V_2'(x)V_1(x) - V_1'(x)V_2(x) = \text{const}/p(x).$$

Thus, the substitution of Equation (7.13) into Equation (7.10) yields

$$Cp(x)W(x)g(x) + C\{[p(x)V_2'(x)]' + q(x)V_2(x)\} \int_a^x V_1(t)g(t)\, dt$$
$$+ C\{[p(x)V_1'(x)]' + q(x)V_1(x)\} \int_x^b V_2(t)g(t)\, dt = g(x).$$

Since $V_1(x)$ and $V_2(x)$ are solutions of the homogeneous equation, this last equation reduces to

$$Cp(x)W(x) \equiv Cp(t)W(t) = 1 \tag{7.14}$$

(since the name of the argument of p and of W does not matter). However, we note that [using the definition of K given in Equation (7.12)]

$$CW(t) = C[V_2'(t)V_1(t) - V_1'(t)V_2(t)] = [K_x(t+, t) - K_x(t-, t)],$$

and it now follows from Equation (7.14) that $K(x, t)$ must have that discontinuity in slope at $x = t$ for which

$$K_x(t+, t) - K_x(t-, t) = 1/p(t). \tag{7.15}$$

Thus, Equation (7.12) defines $K(x, t)$ and Equation (7.15) determines the number C. This function, $K(x, t)$, which we have constructed for use in Equation (7.11) is called the *Green's function* of the problem. It is determined only when the differential operator, the domain, and the homogeneous boundary conditions at the endpoints of that domain have all been specified.

The following problem is illustrative of the foregoing construction.

Let

$$u''(x) - u(x) = f(x) \quad \text{in} \quad 0 < x < 1$$

with

$$u(0) = u(1) = 0.$$

We seek that function, $K(x, t)$, such that

$$u(x) = \int_0^1 K(x, t) f(t) \, dt.$$

Since, according to the foregoing analysis,

$$K(0, t) = u(0) = 0$$

and since $K(x, t)$ in $x < t$ is a solution of $u'' - u = 0$, we have

$$K(x, t) = A \sinh x \quad \text{in} \quad 0 < x < t.$$

We must also require that

$$K(1, t) = 0$$

and that $K(x, t)$ also be a solution of $u'' - u = 0$ in $t < x < 1$. Accordingly, we obtain

$$K(x, t) = B \sinh (1 - x) \quad \text{in} \quad t < x < 1.$$

In order that $K(x, t)$ be continuous at $x = t$, we have

$$A \sinh t = B \sinh (1 - t),$$

that is,

$$K(x, t) = \begin{cases} C \sinh (1 - t) \sinh x, & 0 < x < t; \\ C \sinh (1 - x) \sinh t, & t < x < 1. \end{cases}$$

Finally, since the $p(x)$ of Equation (7.10) in this particular problem is the constant, unity, we have (according to Equation (7.15)), the requirement that

$$K_x(t+, t) - K_x(t-, t) = 1.$$

This says that when

$$-C \cosh (1 - x) \sinh t - C \cosh x \sinh (1 - t)$$

is evaluated at $x = t$, $\quad C$ should be so chosen that

$$-C[\cosh t \sinh (1 - t) + \sinh t \cosh (1 - t)] = 1.$$

A simple manipulation leads to

$$C = -1/\sinh 1.$$

Thus, we have

$$u(x) = \int_0^1 K(x,t) f(t)\, dt$$

$$= -\int_0^x \frac{\sinh(1-x)\sinh t\, f(t)\, dt}{\sinh 1} - \int_x^1 \frac{\sinh x \sinh(1-t) f(t)\, dt}{\sinh 1}.$$

The reader should verify that this function $u(x)$ does satisfy each constraint which was imposed in the statement of the problem. He should also sketch a graph of $K(x,t)$ versus t for each of several values of x.

There is an alternative formalism for the construction of $K(x,t)$ which is very useful and convenient and which is very closely related to the construction in Problem 7.1.4. We define the *delta function*, $\delta(x-t)$ as that function (it really isn't a function in any conventional sense of the word) such that

$$\delta(x-t) = 0 \quad \text{for} \quad x \neq t$$

and

$$\int_{t-\epsilon}^{t+\epsilon} \delta(x-t)\, dx = 1 \quad \text{for any} \quad \epsilon > 0.$$

In a sense, $\delta(x-t)$ may be thought of as a symbolic limit, as $\epsilon \to 0$, for the function $g(x,t)$ defined by

$$g(x,t) = 0, \quad |x-t| > \epsilon; \quad g(x,t) = 1/2\epsilon, \quad |x-t| < \epsilon.$$

We then ask for the solution of the equation

$$(pG'(x,t))' + qG(x,t) = \delta(x-t) \quad \text{in} \quad a < x < b \quad (7.16)$$

with $G(a,t) = G(b,t) = 0$.

It is clear from the definition of $\delta(x-t)$ that the solution $G(x,t)$ of this problem is a solution of the homogeneous equation in $a < x < t$ and in $t < x < b$. We integrate each term of Equation (7.16) from $t-\epsilon$ to $t+\epsilon$ to obtain

$$p(t+\epsilon)G_x(t+\epsilon,t) - p(t-\epsilon)G_x(t-\epsilon,t) + \int_{t-\epsilon}^{t+\epsilon} q(x)G(x,t)\, dx = 1,$$

$$(7.17)$$

and, in the limit as $\epsilon \to 0$, we obtain the requirement that

$$G_x(t+,t) - G_x(t-,t) = 1/p(t). \quad (7.18)$$

Since the foregoing statements are merely a repetition of the criteria which define $K(x,t)$, we see that $G(x,t)$ is precisely the $K(x,t)$ we found above.

We now write Equations (7.16) and (7.10) in the form

$$u(x)\{[pG'(x, t)]' + qG(x, t)\} = \delta(x - t)u(x)$$

and

$$G(x, t)\{[pu'(x)]' + qu(x)\} = g(x)G(x, t),$$

where we seek that solution of Equation (7.10) for which $u(a) = u(b) = 0$. Subtracting these two equations, we obtain

$$[p(uG' - Gu')]' = \delta(x - t)u(x) - g(x)G(x, t).$$

Integration over x from a to b then gives

$$u(t) = \int_a^b g(x)G(x, t)\,dx.$$

Since, using Equation (7.12), $G(x, t) = K(x, t) = K(t, x)$, we have recovered Equation (7.11).

In our *Green's function* discussion, as well as in the previous discussion of Wronskians, we have usually written differential operators in the form

$$[p(x)u']' + q(x)u \tag{7.19}$$

rather than in the form

$$L_1(u) = a_2(x)u'' + a_1(x)u' + a_0(x)u. \tag{7.20}$$

The reason for this is that the "multiply and subtract" process is easier for the form (7.19) than for the form (7.20). If we want to use this process for the form (7.20), it is convenient to introduce an *adjoint operator* L_2 defined by

$$L_2(u) = (a_2 u)'' - (a_1 u)' + a_0 u. \tag{7.21}$$

The "multiplication and subtraction" process then leads to

$$\int_a^b [uL_1(v) - vL_2(u)]\,dx = [a_2(uv' - vu') + (a_1 - a_2')uv]_a^b, \tag{7.22}$$

where u, v are any differentiable functions. If $L_2 \equiv L_1$ (that is, if $a_1 = a_2'$), the operator L_1 is said to be *self-adjoint*; notice the simplification in Equation (7.22) for this case. Observe also that, if $a_1 = a_2'$, then L_1 has the form (7.19).

7.3 Problems

7.3.1 (a) Let $G(x, t)$ satisfy Equation (7.16), and let $G(x, \xi)$ satisfy Equation (7.16) with t replaced by ξ. Use the "multiply and subtract" technique on these two equations to show that

$G(x, t) = G(t, x)$; that is, show that the Green's function for a self-adjoint operator is *symmetric*.

(b) Let $K(x, t)$ be the Green's function associated with

$$L_1(w) = \delta(x - t) \quad \text{in} \quad a < x < b,$$

with $w(a) = w(b) = 0$, and let $H(x, t)$ be the Green's function for

$$L_2(v) = \delta(x - t) \quad \text{in} \quad a < x < b$$

with $v(a) = v(b) = 0$.

In these equations, L_1 and L_2 are the operators of Equations (7.20) and (7.21).

Find H in terms of K. Find $\phi(x)$ in terms of $K(x, t)$, where $L_1(\phi) = r(x)$ in $a < x < b$, with $\phi(a) = \phi(b) = 0$.

7.3.2 Find, sketch, and compare the Green's function of

$$u'' + \sigma^2 u = h(x) \quad \text{in} \quad 0 < x < \pi$$

with $u(0) = u(\pi) = 0$ and the Green's function of

$$u'' - \sigma^2 u = h(x) \quad \text{in} \quad 0 < x < \pi$$

with $u(0) = u(\pi) = 0$.

In each of these, include both large and small values of σ. Using these Green's functions, find $u(x)$ for the particular problem in which $h(x) = \sin x$. Are there any values of the constant σ which give trouble? Compare the results with those of Problem 7.1.1.

7.3.3 Find, and sketch graphs of, the Green's functions associated with the following problems:

(a) $W''(x) = f(x)$ in $0 < x < 1$ with $W(0) = W(1) = 0$;

(b) $x(xV'(x))' - \nu^2 V(x) = g(x)$ in $0 < x < 1$ with $V(0) = V(1) = 0$; ν is a positive constant;

(c) $x(xy'(x))' - \nu^2 y(x) = h(x)$ in $1 < x < a$ with $y(1) = y'(a) = 0$;

(d) $x(xz'(x))' - \nu^2 z(x) = R(x)$ in $0 < x < 1$ with $z'(0) = z(1) = 0$.

Are there any positive values of ν which are particularly troublesome in problems (b), (c), (d)?

In the graphs you draw, use both small and large values of v, deciding for yourselves the appropriate meanings of "large" and "small."

7.3.4 Generalize the result of Problem 7.1.7 so as to apply to the case in which the equation $L_2(y) = 0$, with $y(a) = y(b) = 0$, has a nontrivial solution, and in which it is desired to solve the equation $L_1(y) = g(x)$ with $y(a) = y(b) = 0$. Does your result imply that the Green's function method may not always work? Can you modify it for troublesome cases? (This is hard.)

7.3.5 Let

$$(xu')' - \frac{9}{x}u = H(x)$$

in

$$0 < x < 10, \quad \text{with} \quad u(0) = u(10) = 0.$$

(a) Find the Green's function, $G(x, t)$, associated with this equation. Plot $G(x, t)$ against t. Use $G(x, t)$ to write u as an integral.

(b) Study (using graphs of G and N) the possibility of approximating $G(x, t)$ by an "approximate kernel" $N(x, t)$ of the form

$$N(x, t) = A(x) \begin{cases} e^{-\alpha(x-t)}, & x > t, \\ e^{-\alpha(t-x)}, & x < t. \end{cases}$$

What should $A(x)$ and α be?

(c) Use each of G and N to find u when $H(x) = \cos \beta x$, for each of the cases $\beta = .3, \beta = 3$.

(d) Enunciate some guidelines for the use of approximate kernels.

7.3.6 Consider the eigenvalue problem

$$(py')' + (q + \lambda r)y = 0 \quad \text{(homogeneous boundary conditions)}.$$

Rewrite this in the form

$$(py')' + qy = -\lambda ry.$$

Treating $-\lambda ry$ as a nonhomogeneous term, write the solution of this equation in terms of the Green's function associated with the equation

$$(py')' + qy = f(x)$$

and the same boundary conditions. The result is called an *integral equation*. Do we still have to include the boundary conditions when the eigenvalue problem is stated in integral equation form?

7.4 Nonhomogeneous Boundary Conditions

The foregoing discussion of Green's functions has been concerned with solutions of Equation (7.1) with homogeneous boundary conditions.

§7.4] Nonhomogeneous Boundary Conditions

Frequently, of course, one encounters a mathematical problem in which he seeks the solution of Equation (7.1) with nonhomogeneous boundary conditions. For example, let

$$F''(x) + xF(x) = e^{3x}, \quad \text{in} \quad 0 < x < 1, \tag{7.23}$$

with

$$F(0) = 0, \quad F(1) = 3. \tag{7.24}$$

If we wish to use the foregoing development, we may proceed as follows. Define

$$H(x) = F(x) + M(x),$$

where the function $M(x)$ is to satisfy certain boundary conditions; apart from this, its selection is only a matter of convenience. In particular, we demand of $M(x)$ only that $M(0) = 0$, that $M(1) = -3$, and that $M(x)$ be twice differentiable in $0 < x < 1$. If we call $M''(x) + xM(x) = P(x)$, Equations (7.23) and (7.24) become

$$H''(x) + xH(x) = e^{3x} + P(x) = R(x)$$

with

$$H(0) = H(1) = 0.$$

The determination of $H(x)$ falls within the scope of the foregoing Green's function treatment, and H can be described by Equation (7.11) with the Green's function which can be constructed using Equations (4.4) and (4.5).

One obvious possible choice for $M(x)$ is

$$M(x) = -3x.$$

With that choice, we obtain

$$F(x) = 3x + \int_0^1 K(x,t)(e^{3t} - 3t^2)dt,$$

where K is the appropriate Green's function.

The Green's function also plays a very valuable role in the treatment of partial differential equations. A discussion of the important techniques (and more on the delta function) can be found in G. Carrier and C. Pearson, *Partial Differential Equations*, 2nd edition, Academic Press, 1988.

Expansions in Eigenfunctions | 8

Let
$$[p(x)\chi'(x)]' + q(x)\chi(x) = h(x) \tag{8.1}$$
in $a < x < b$, with
$$\chi(a) = \chi(b) = 0.$$

We have seen that, when p and q are reasonably well behaved functions, $\chi(x)$ can be written as an integral like that in Equation (7.13). There are circumstances under which an alternative description is desirable; one useful alternative is that in which χ is described as a series in the eigenfunctions of an appropriate Sturm-Liouville problem. The choice of the particular Sturm-Liouville problem depends on the subsequent use to which the description of $\chi(x)$ is to be put, and one cannot give a firm rule for such a choice. Nevertheless, it is usually most useful to ask for a description of $\chi(x)$ in terms of the eigenfunctions of
$$[p(x)\phi'(x)]' + q(x)\phi(x) + \lambda r(x)\phi(x) = 0 \tag{8.2}$$
with $\phi(a) = \phi(b) = 0$.

In Equation (8.2), $r(x)$ may be the same as $q(x)$ or it may be a choice dictated by features of the specific question under study. In either event, with $p(x)$ and $r(x)$ positive in $a \leq x \leq b$, we denote the normalized eigenfunctions by $\phi_n(x)$, the eigenvalues by λ_n, and we write the expected description of χ in the form

$$\chi(x) = \sum A_n \phi_n(x), \tag{8.3}$$

where the equality is expected to hold in the mean-square sense.

We now multiply Equation (8.1) by ϕ_k to obtain

$$\phi_k[(p\chi')' + q\chi] = \phi_k h$$

and, from Equation (8.2),

$$\chi[(p\phi_k')' + q\phi_k + \lambda_k r \phi_k] = 0.$$

Subtracting one from the other, we obtain

$$[p(\phi_k \chi' - \chi \phi_k')]' - \lambda_k r \phi_k \chi = \phi_k h.$$

Using the boundary conditions, we integrate over $a < x < b$ to obtain

$$\lambda_k \int_a^b r \phi_k \chi \, dx = -\int_a^b \phi_k h \, dx. \tag{8.4}$$

Using Equations (6.13) and (8.3), we identify the left-hand side of Equation (8.4) as $\lambda_k A_k$ and, using Equation (6.13) again, we note that the right side is the expansion coefficient of $-h(x)/r(x)$. Thus, the completeness of the set of eigenfunctions (which we have asserted but not demonstrated) assures us that

$$\chi(x) = \sum A_n \phi_n(x)$$

with

$$A_k = -\left[\int_a^b \phi_k(x) h(x) \, dx\right] \Big/ \lambda_k. \tag{8.5}$$

This simple relation between the expansion coefficients of $h(x)$ and $\chi(x)$ is noteworthy.

8.1 Problems

8.1.1 Re-examine Problems 7.1.3 and 7.3.2. Pursue each, using eigenfunction descriptions of the function sought. Comment on the relative merits of the three techniques (use of the Wronskian, Green's function method, or eigenfunction expansion) for each problem, and for particular values of the parameters of the problems.

8.1.2 (a) Modify the analysis of the preceding section, identified with a problem in which the boundary conditions are $u(a) = u(b) = 0$, so that the result is generalized to apply under any homogenous boundary conditions.

(b) In Equation (8.1), we required $\chi(a) = \chi(b) = 0$. Suppose instead that $\chi(a) = A$ and $\chi(b) = B$; carry out the corresponding analysis, and obtain the equation which replaces Equation (8.5). (Do not use the device of part (c) below, but use the "multiply and subtract" technique on Equations (8.1) and (8.2).)

(c) Under the discussion of Green's functions at the end of Chapter 7, we noted that a rudimentary device enabled us to convert a problem with nonhomogeneous boundary conditions to one with homogeneous boundary conditions. Can you see any advantage in using and/or modifying this device when describing the solution of such a problem by an eigenfunction expansion?

8.1.3 Use the techniques of each of parts (b) and (c) of Problem (8.1.2) to obtain an expansion for $f(x) = x^2$, in $0 < x < \pi$, in terms of the eigenfunctions of the equation $\phi'' + \lambda^2 \phi = 0$, $\phi(0) = \phi(\pi) = 0$. For the differential equation satisfied by $f(x)$, use $f'' = 2$, and in using the method of part (c), set $f = y(x) + \pi x$.

8.1.4 You have had an opportunity under each of several recent problems to infer an important fact concerning the existence of solutions of nonhomogeneous equations with homogeneous boundary conditions.

Equation (8.4), in particular, indicates clearly that the solution of Equation (8.1) under the given boundary conditions *may* fail to exist if one of the eigenvalues, say λ_N, of Equation (8.2) has the value zero. Show that, with such an eigenvalue, it *does* fail to exist unless

$$\int_a^b \phi_N(x)h(x)\,dx = 0.$$

Show also that, if this condition is satisfied, a solution does exist. (Assume that term-by-term differentiation of Equation (8.3) is valid.) Is the solution unique?

8.1.5 Does the lack of existence discussed in Problem 8.1.4 carry over into the problem in which, for example, $\psi(a) = 0$, $\psi(b) = 1$?

8.1.6 Obtain an eigenfunction expansion for the solution of

$$(xy')' + \frac{3y}{x} = \sin x,$$

$y(1) = y(2) = 0$, in $1 \leq x \leq 2$. Use the eigenfunctions associated with

$$(xy')' + \frac{3+\lambda}{x}y = 0,$$

$y(1) = y(2) = 0$.

8.2 Two Approximation Methods

Eigenfunction expansions are also useful in the discussion of approximation techniques. We illustrate this by considering two examples concerned with the approximate determination of the lowest eigenvalue λ_1 of the problem of Equation (8.2) for the case $\lambda_1 > 0$. As before, denote the eigenvalues by $\lambda_1, \lambda_2, \lambda_3, \cdots$ (with $\lambda_1 < \lambda_2 < \lambda_3 < \cdots$), and the corresponding eigenfunctions by $\phi_1, \phi_2, \phi_3, \cdots$.

In our first example, we start with some chosen function $y_0(x)$, satisfying $y_0(a) = y_0(b) = 0$ and having the eigenfunction expansion

$$y_0(x) = \sum_{k=1}^{\infty} B_k \phi_k(x), \tag{8.6}$$

where $B_k = \int_a^b r y_0 \phi_k \, dx$ as before. If we now solve the problem

$$(py_1')' + qy_1 = -ry_0, \tag{8.7}$$

$y_1(a) = y_1(b) = 0$, then Equation (8.5) requires that

$$y_1(x) = \sum_{k=1}^{\infty} \frac{B_k}{\lambda_k} \phi_k(x).$$

Similarly, the solution $y_2(x)$ of

$$(py_2')' + qy_2 = -ry_1, \tag{8.8}$$

$y_2(a) = y_2(b) = 0$, is given by

$$y_2(x) = \sum_{k=1}^{\infty} \frac{B_k}{\lambda_k^2} \phi_k(x),$$

and continuation shows that the solution of

$$(py_n')' + qy_n = -ry_{n-1}, \tag{8.9}$$

$y_n(a) = y_n(b) = 0$ will be given by

$$y_n(x) = \sum_{k=1}^{\infty} \frac{B_k}{\lambda_k^n} \phi_k(x). \tag{8.10}$$

Since λ_1 is the lowest eigenvalue, the first term in this sum will tend to dominate the other terms as n becomes large (unless y_0 is such that $B_1 \equiv 0$). Thus, as $n \to \infty$, we have

$$y_n \sim \frac{B_1}{\lambda_1^n} \phi_1,$$

and it follows that

$$\lim_{n \to \infty} \frac{y_n}{y_{n+1}} = \lambda_1. \tag{8.11}$$

The approximation technique based on this result consists in choosing some simple function y_0 arbitrarily, and then solving in succession Equations (8.7), (8.8), ..., so as to give $y_1, y_2, \ldots$. We then stop at y_{N+1}, say, and use the value of the ratio of y_N to y_{N+1} as an approximation to λ_1. This ratio will, of course, be a function of x, so that we must average it in some reasonable manner over the whole interval (if N is large enough, the dependence of the ratio on x will be unimportant). As a bonus, the function $y_{N+1}(x)$ is an approximation to a multiple of the first eigenfunction $\phi_1(x)$.

As a second example, let $w(x)$ be some chosen function, satisfying $w(a) = w(b) = 0$, and having the expansion

$$w(x) = \sum_{k=1}^{\infty} C_k \phi_k(x). \tag{8.12}$$

We construct the *Rayleigh quotient*

$$R = \frac{\int_a^b [p(w')^2 - qw^2]\, dx}{\int_a^b rw^2\, dx} \tag{8.13}$$

(already encountered in another connection in Problem 6.1.6) and substitute the right-hand side of Equation (8.12) for w to obtain

$$R = \frac{\int_a^b [p(\sum C_k \phi_k')^2 - q(\sum C_k \phi_k)^2]\, dx}{\int_a^b r(\sum C_k \phi_k)^2\, dx}.$$

The first term in the numerator may be integrated by parts to give

$$[p(\sum C_k \phi_k') \sum C_j \phi_j]_a^b - \int_a^b \sum C_k (p\phi_k')' \cdot \sum C_j \phi_j\, dx,$$

and use of the condition $\phi_j(a) = \phi_j(b)$ and of the differential equation satisfied by the ϕ_k now leads to

$$R = \frac{\int_a^b (\sum C_k \lambda_k r \phi_k)(\sum C_j \phi_j)\, dx}{\int_a^b r(\sum C_k \phi_k)^2\, dx}.$$

Since the ϕ_j are orthonormal, this expression reduces to

$$R = \frac{\lambda_1 C_1^2 + \lambda_2 C_2^2 + \lambda_3 C_3^2 + \cdots}{C_1^2 + C_2^2 + C_3^2 + \cdots}. \tag{8.14}$$

The reader should now convince himself that, as we consider a number of different functions $w(x)$ with $w(a) = w(b) = 0$ (each such function involving its own set of C_j's), the value of R is always $\geq \lambda_1$, and this minimal value is attained if w is a multiple of ϕ_1. Thus, we can determine λ_1 approximately by trying various functions w in Equation (8.13), and setting λ_1 equal to the least value of R so obtained.

8.3 Problems

8.3.1 Extend the discussion of approximation techniques to the case in which the homogeneous boundary conditions are of a more general character.

8.3.2 Carry through the sequence of calculations (8.7), (8.8), ..., in order to determine (approximately) the lowest eigenvalue and corresponding eigenfunction for each of

(a) $\qquad y'' + \lambda y = 0, \qquad y(0) = y(\pi) = 0,$

(b) $\qquad y'' + \lambda x y = 0, \qquad y(0) = y(\pi) = 0.$

Compare your results with the exact solutions.

8.3.3 We remarked, following Equation (8.11), that the ratio obtained by stopping after N iterations, namely, $y_N(x)/y_{N+1}(x)$, would be a function of x and so would have to be averaged in some manner. Discuss the relative efficiency of such averaging formulas as

(a) (b)

$$\frac{\int_a^b y_N(x)\,dx}{\int_a^b y_{N+1}(x)\,dx}, \qquad \frac{\int_a^b r(x) y_N^2(x)\,dx}{\int_a^b r(x) y_{N+1}(x) y_N(x)\,dx},$$

(c) (d)

$$\frac{\int_a^b |y_N(x)|\,dx}{\int_a^b |y_{N+1}(x)|\,dx}, \qquad \left[\frac{\int_a^b r(x) y_N^2(x)\,dx}{\int_a^b r(x) y_{N+1}^2(x)\,dx}\right]^{1/2}.$$

8.3.4 Noting the parenthetical remark following Equation (8.10), devise an approximation technique for finding higher eigenvalues, and use it to find λ_2 and ϕ_2 for the two examples of Problem 8.3.2. Describe a similar procedure for the Rayleigh quotient (8.13).

8.3.5 One method for obtaining an approximate least value for the Rayleigh quotient (8.13) consists in writing

$$w = \alpha_1 w_1 + \alpha_2 w_2 + \cdots + \alpha_N w_N,$$

where $w_1(x), \ldots, w_N(x)$ are N chosen functions, and $\alpha_1, \ldots \alpha_N$ are N undetermined constants. (The boundary conditions $w(a) = w(b) = 0$

must be satisfied for any choice of the α_j.) If this expression for w is then substituted into Equation (8.13), the constants α_j may be adjusted so as to minimize R. Carry out this *Rayleigh–Ritz* procedure for the two problems of Problem 8.3.2. For what set of functions $w_j(x)$ could this procedure be guaranteed to yield λ_1, exactly, as $N \rightarrow \infty$?

The Perturbation Expansion | 9

Frequently, in the foregoing text and exercises, we have found that the solution $u(x, \alpha)$ of a differential equation containing a parameter, α, in one of its coefficients is a differentiable function of that parameter. (For example, the solutions, $\cos(x\sqrt{\alpha})$ and $\sqrt{\alpha}\sin(x\sqrt{\alpha})$ of Equation (3.8), have derivatives with regard to α for all α, and in Equations (4.28) and (4.29) we have already exploited the derivative with regard to β of a solution of Equation (4.24).) Accordingly, one might hope that the function $u(x, \alpha)$ would admit a description

$$u(x, \alpha) = u_0(x) + \alpha u_1(x) + \alpha^2 u_2(x) + \cdots$$

and that the use of such a description for small enough values of α might provide a very useful tool. As an illustrative example of this sort, we study the problem (with primes denoting differentiation with regard to x)

$$u''(x, \alpha) - (1 + \alpha x e^{-2x})u(x, \alpha) = 0, \tag{9.1}$$

in $x > 0$, with $u(0, \alpha) = 1$, $u'(0, \alpha) = 1$, and $0 < \alpha \ll 1$. We ask whether $u(x, \alpha)$ can be described by a convergent series

$$u(x, \alpha) = \sum_{n=0}^{\infty} u_n(x) \alpha^n. \tag{9.2}$$

If we substitute Equation (9.2) into Equation (9.1), we obtain

$$(u_0'' - u_0) + \alpha(u_1'' - u_1 - xe^{-2x}u_0) + \alpha^2(u_2'' - u_2 - xe^{-2x}u_1) + \cdots = 0, \tag{9.3}$$

and, if the series described in Equation (9.3) is to converge to zero for each sufficiently small value of α, the coefficient of each power of α must vanish. Furthermore, since we require that $u(0, \alpha) = 1$ and $u'(0, \alpha) = 1$, we have

$$\sum_{n=0}^{\infty} u_n(0)\alpha^n = 1 \quad \text{and} \quad \sum_{n=0}^{\infty} u'_n(0)\alpha^n = 1,$$

so that

$$u_0(0) = 1, \quad u_n(0) = 0, \quad n > 0; \quad u'_0(0) = 1, \quad u'_n(0) = 0, \quad n > 0. \quad (9.4)$$

According to Equations (9.3) and (9.4), $u''_0 - u_0 = 0$ with $u_0(0) = u'_0(0) = 1$, and it follows that

$$u_0 = e^x. \quad (9.5)$$

Using Equations (9.3) and (9.4) again, we have

$$u''_1 - u_1 = xe^{-x} \quad (9.6)$$

with

$$u_1(0) = u'_1(0) = 0.$$

It is easy to verify (or to deduce by methods discussed earlier) that

$$u_1(x) = \tfrac{1}{4}[\sinh x - (x^2 + x)e^{-x}]. \quad (9.7)$$

A continuation of the process yields

$$u_2 = -\tfrac{1}{256}[677 \sinh x + \{16(x^2 + x)\cosh x - 752 \sinh x$$
$$+ (8x^3 + 18x^2 + 59x)e^{-x}\}e^{-2x}],$$

$$u_3 = \tfrac{1}{256}\Big[-\tfrac{3787}{96}\sinh x + \{\tfrac{667}{4}(x^2 + x)\sinh x + \tfrac{11189}{288}e^x\}e^{-2x}$$
$$+ \tfrac{1}{32}\{64x^3 + 5544x^2 + 124x - 3935\}\cosh x e^{-4x}$$
$$+ \left\{\tfrac{x^4}{3} + \tfrac{23}{36}x^3 - \tfrac{657}{8}x^2 + \tfrac{113}{36}x + \tfrac{12113}{144}\right\}e^{-5x}\Big].$$

Not only could we prove that the series (9.2) converges* for all α, but more importantly in regard to the utility of the method, we see that a very few terms suffice to describe $u(x, \alpha)$ accurately for all $x > 0$ and $\alpha \ll 1$.

A rather different situation prevails when we look at the superficially similar problem in which

---

* A more general proof can be found in Burkill, *The Theory of Ordinary Differential Equations*, Oliver and Boyd, Ltd., 1956.

$$v''(x, \alpha) - (1 + \alpha x e^{2x})v(x, \alpha) = 0 \tag{9.8}$$

and

$$v(0, \alpha) = 1, \quad v'(0, \alpha) = 1.$$

The reader should verify that the same reasoning and procedure lead to

$$v_0 = e^x,$$

$$v_1 = \tfrac{1}{16}[\sinh x + \{2xe^x - 3 \sinh x\}e^{2x}],$$

$$v_2 = \frac{1}{256}\left[\left(-2x^2 + 2x + \frac{7}{108}\right)e^x + (4x - 3)e^{3x}\right.$$
$$\left. + \left(\frac{4x^2}{3} - \frac{19}{9}x + \frac{89}{108}\right)e^{5x} + \frac{19}{9}\cosh x\right].$$

Although the series for $v(x, \alpha)$ also converges for all α, it is abundantly clear from inspection of the growth with x of each $v_j(x)$ that many terms of the series

$$v(x, \alpha) = \sum_{j=0}^{\infty} v_j(x)\alpha^j \tag{9.9}$$

would be required to provide a useful description over any appreciable range of x.

The cause of the discrepancy between the results of the two problems is not deeply hidden. One can hope that this method will provide an efficient description only if the term multiplied by α plays the role of a small correction. That is, if, in Equation (9.1), $u''(x, \alpha)$ is almost equal to $u(x, \alpha)$ for all x of interest, so that $u(x, \alpha) \cong e^x$, then the small correction implied by the remaining term is supplied iteratively by the successive corrections, $\alpha u_1(x)$, $\alpha^2 u_2(x)$, etc. On the other hand, in Equation (9.8), $\alpha x e^{2x} v(x, \alpha)$ will be much larger than $v(x, \alpha)$ for all $x \gg -\ln \alpha$. Therefore, the terms of Equation (9.8) which must nearly balance for all such x are $v''(x, \alpha)$ and $\alpha x e^{2x} v(x, \alpha)$. For such x, the first term of the series in Equation (9.9) cannot be a good approximation to $v(x, \alpha)$ and the subsequent terms cannot be merely small corrections. Hence, the method fails to be useful in any direct way.*

Thus, the perturbation series, as such, cannot provide a useful descrip-

---

* It is very important not to overlook the existence of a technique which sometimes makes a series like (9.9) very valuable. It may be possible to rearrange the terms of such a series and deduce, analytically, the sum of the dominant contributions. Such techniques are beyond the scope of this course, but the reader should not write off the possibilities inherent in a series like (9.9).

tion of a solution of $L(u) + \alpha L_1(u) = 0$ unless $\alpha L_1(u)$ is uniformly small compared to the individual terms which make up $L(u)$; here L and L_1 denote differential operators. Furthermore, this is a necessary condition, not a sufficient one. We shall encounter problems later in which nonuniformities in the behavior of such series arise in more subtle ways.

9.1 Problems

9.1.1 Use a perturbation method to find two linearly independent solutions of

$$(xu')' + (\alpha^2 x - \nu^2/x)u = 0,$$

where

$$\alpha \ll 1.$$

Compare the results with those of Problem 4.3.1.

9.1.2 With the results of Problem (7.1.2), use a perturbation technique to find, when $|\alpha - 3| \ll 1$, the solution of

$$(xu')' - (x - \alpha)u = 0$$

for which $u(0) = 1$.

9.1.3 Estimate a value of α such that

$$(xu')' - (x - \alpha)u = 0 \quad \text{in} \quad 0 < x < 1$$

with

$$u(0) = 1 \quad \text{and} \quad u(1) = 0.$$

Estimate a value for β for which

$$(xu')' - (x - \beta)u = 0 \quad \text{in} \quad 1 < x < \infty$$

with

$$u(1) = 0 \quad \text{and} \quad u(\infty) = 0.$$

9.2 An Eigenvalue Problem

Consider now the eigenvalue problem in which

$$u''(x, \alpha) - \alpha x u(x, \alpha) + \lambda u(x, \alpha) = 0 \qquad (9.10)$$

in $0 < x < \pi$, with the given parameter $0 < \alpha \ll 1$ and

$$u(0, \alpha) = u(\pi, \alpha) = 0. \qquad (9.11)$$

It is evident that any eigenvalue λ_n will be an increasing function of α (the larger α is, the larger λ_n must be in order to provide enough positive coefficient on u to pull u back to zero as quickly as it did when α was smaller).

For small values of α, we can attempt to describe any eigenfunction $u_n(x, \alpha)$ as a power series in α; however, since the corresponding eigenvalue λ_n is also a function of α, we must anticipate the descriptions

$$\lambda_n(\alpha) = \sum_{j=0}^{\infty} \lambda_{nj} \alpha^j \quad \text{and} \quad u_n(x, \alpha) = \sum_{j=0}^{\infty} u_{nj}(x) \alpha^j.$$

We confine our attention first to the lowest eigenvalue λ_1 and to $u_1(x)$, and we write

$$\lambda_1(\alpha) = \sum_{j=0}^{\infty} \sigma_j \alpha^j, \tag{9.12}$$

$$u_1(x, \alpha) = \sum_{j=0}^{\infty} w_j(x) \alpha^j \tag{9.13}$$

just in order to avoid the double subscripts. If we substitute Equations (9.12) and (9.13) into Equation (9.10), we obtain

$$(w_0'' + \sigma_0 w_0) + \alpha(w_1'' + \sigma_0 w_1 + \sigma_1 w_0 - x w_0)$$
$$+ \alpha^2(w_2'' + \sigma_0 w_2 + \sigma_1 w_1 + \sigma_2 w_0 - x w_1) + \cdots = 0 \tag{9.14}$$

and, if we substitute Equation (9.13) into Equation (9.11), we obtain

$$\sum_{j=0}^{\infty} w_j(0) \alpha^j = \sum_{j=0}^{\infty} w_j(\pi) \alpha^j = 0. \tag{9.15}$$

These results imply that

$$w_0''(x) + \sigma_0 w_0(x) = 0 \quad \text{with} \quad w_0(0) = w_0(\pi) = 0,$$

so that we find

$$w_0(x) = \sin x \tag{9.16}$$

and

$$\sigma_0 = 1. \tag{9.17}$$

We then find from Equations (9.14) and (9.15) that

$$w_1'' + w_1 = x w_0 - \sigma_1 w_0 \tag{9.18}$$

with

$$w_1(0) = w_1(\pi) = 0. \tag{9.19}$$

The general solution of Equation (9.18) is

$$w_1(x) = \frac{x \sin x - x^2 \cos x}{4} + \frac{\sigma_1}{2} x \cos x + A \sin x + B \cos x.$$

In order that $w_1(0) = 0$, we must choose $B = 0$, and the condition $w_1(\pi) = 0$ reduces to

$$w_1(\pi) = 0 = \frac{\pi^2}{4} - \frac{\pi \sigma_1}{2},$$

so that we have

$$\sigma_1 = \pi/2. \qquad (9.20)$$

An alternative procedure would have been to observe that Equation (9.18) cannot have a solution unless the nonhomogeneous term is orthogonal to the function $\sin x$ (compare with Problem 8.1.4); this condition leads again to Equation (9.20).

Thus, the first two terms of the perturbation expansion give

$$\lambda_1 = 1 + \alpha \frac{\pi}{2},$$

$$u_1 = \sin x + \alpha \left[\frac{x \sin x - x^2 \cos x}{4} + \frac{\pi}{4} x \cos x \right] + \alpha A \sin x. \qquad (9.21)$$

The constant A is as yet undetermined. If we leave A as arbitrary for the moment, then the reader may verify that w_2 will contain the term

$$A \left[\frac{x \sin x - x^2 \cos x}{4} + \frac{\pi}{4} x \cos x \right],$$

so that the combination $w_0 + \alpha w_1 + \alpha^2 w_2$ will contain the term

$$\alpha A \left[\sin x + \alpha \left\{ \frac{x \sin x - x^2 \cos x}{4} + \frac{\pi}{4} x \cos x \right\} \right].$$

Continuation of the process shows that the terms involving A provide merely a multiple of the original perturbation series of Equation (9.21). Thus, the first two terms of Equation (9.21) are replaced by

$$w = (1 + \alpha A) \left\{ \sin x + \alpha \left[\frac{x \sin x - x^2 \cos x}{4} + \frac{\pi}{4} x \cos x \right] \right\}.$$

However, the eigenfunction is undetermined within a multiplicative constant in any event; consequently, there is no loss in generality in simply setting $A = 0$. Similarly, we set equal to zero any constant whose value is not determined by the perturbation process. Thus, $A = 0$ and Equation (9.21) becomes

$$u_1 = \sin x + \alpha \left[\frac{x \sin x - x^2 \cos x}{4} + \frac{\pi}{4} x \cos x \right] + 0(\alpha^2), \quad (9.22)$$

where the expression $0(\alpha^2)$ means a term which vanishes in proportion to α^2 as $\alpha \to 0$.

Clearly, the process can be continued and one can estimate that term of the series for λ_1 and u_1 at which termination of the series seems to provide a reasonable compromise between the wish for accuracy and the availability of time and effort. Only if α is small enough, of course, will this approach provide an effective description of the answer to our question.

9.3 Problems

9.3.1 Find $w_2(x)$, and σ_2, for the foregoing problem.

9.3.2 Solve the problem again, normalizing $u_1(x, \alpha)$ according to the rule

$$u'(0, \alpha) = 1.$$

Show that (to the extent to which you have found λ_1 and u_1) the value of λ_1 so obtained has precisely the same series expansion as that of Problem 9.3.1. Show, also, that the series for $u_1(x, \alpha)$ so obtained can be written as $f(\alpha)$ times the $u_1(\alpha)$ found in the text. Here $f(\alpha)$ is a power series of the form

$$f(\alpha) = \sum_{n=0}^{\infty} f_n \alpha^n,$$

where the f_n are just numbers.

9.3.3 Can you use the same process on the problem

$$\alpha u''(x, \alpha) + (x - \sin x) u(x, \alpha) = 0$$

with

$$u(0) = u(\pi) = 0$$

and with $\alpha \ll 1$? Why?

Asymptotic Series | 10

In the foregoing text we have represented the solutions of differential equations by a series in powers of the independent variable, by a series in powers of a parameter of the problem, and by a series of eigenfunctions. Another, rather different, representation which is also of great value makes use of *asymptotic series*. Before giving a careful definition of an asymptotic series, we consider the illustrative question: how does the *exponential integral function*

$$Ei(x) = \int_x^\infty \frac{e^{-s}}{s} ds \qquad (10.1)$$

behave for large positive values of x?

Having no better idea at the moment, we integrate by parts to obtain

$$Ei(x) = \frac{e^{-x}}{x} - \int_x^\infty \frac{e^{-s}}{s^2} ds = \frac{e^{-x}}{x} - \frac{e^{-x}}{x^2} + 2\int_x^\infty \frac{e^{-s}}{s^3} ds = \cdots$$

$$= \frac{e^{-x}}{x}\left[1 - \frac{1}{x} + \frac{2!}{x^2} + \cdots + \frac{(-1)^n n!}{x^n}\right] + R_n$$

$$= \frac{e^{-x}}{x} \sum_{j=0}^n a_j x^{-j} + R_n, \qquad (10.2)$$

where

$$R_n = (-1)^{n+1}(n+1)! \int_x^\infty \frac{e^{-t}}{t^{n+2}} dt. \qquad (10.3)$$

We see that

Asymptotic Series

$$|R_n| \leq (n+1)! \int_x^\infty \frac{e^{-t}}{x^{n+2}}\, dt = (n+1)!\, \frac{e^{-x}}{x^{n+2}} \tag{10.4}$$

and also that $|a_{n+1}x^{-(n+2)}| > |a_n x^{-(n+1)}|$ whenever $n > x$. Thus, the series

$$\sum_{n=0}^\infty a_n x^{-n}$$

does not converge for any x and cannot be said to have a sum in the conventional sense. However, $R_n(x)$ does tend to zero as $x \to \infty$ for fixed values of n.

Furthermore, for large x (say $x = 10$), $|R_4(10)| \simeq e^{-10}\, 10^{-4}$, whereas

$$Ei(10) = e^{-10}(.0914) + R_4,$$

and the error incurred when $Ei(10)$ is calculated from the approximation

$$Ei(x) \cong \frac{e^{-x}}{x} \sum_{n=0}^4 a_n x^{-n} \tag{10.5}$$

is only about one in 10^3. Thus, one can anticipate that, even though the series

$$Ei(x) \sim \frac{e^{-x}}{x} \sum_{n=0}^\infty a_n x^{-n} \tag{10.6}$$

does not converge, it may serve a very useful purpose. One says, in fact, that this useful series is an *asymptotic representation* for $Ei(x)$.

More generally, whenever we can write

$$F(x) = \sum_{n=0}^N b_n x^{-n} + R_N(x)$$

and establish that

$$x^N R_N(x) \to 0 \quad \text{as} \quad x \to \infty,$$

the series $\sum_{n=0}^\infty b_n x^{-n}$ is said to be asymptotic to $F(x)$ and this relationship is written * as

$$F(x) \sim \sum_{n=0}^\infty b_n x^{-n}.$$

---

* This definition frequently is extended to include more exotic series. For example, one series encountered in a practical problem has the form
$$G(x) \sim a_0 + a_1 x^{-2/3} + a_2 x^{-1} \ln x + a_3 x^{-1} + a_4 x^{-4/3} + \cdots.$$
In this book, we will not need any series more general than those containing powers of x^{-1}, but it is worth noting that one can
 (1) define as an *asymptotic sequence* of functions, $\phi_n(x)$, any set of functions such that $\phi_{k+1}(x)/\phi_k(x) \to 0$ as $x \to \infty$;
 (2) define as an *asymptotic series* for $f(x)$, any representation $f(x) = \sum_{k=0}^n a_k \phi_k(x) + R_n$ for which $R_n(x)/\phi_n(x) \to 0$ as $x \to \infty$ for each integer n.

To use this language meticulously on the foregoing example, we must write

$$G(x) = xe^x \, Ei(x) \sim \sum_{n=0}^{\infty} a_n x^{-n},$$

and this is the representation of $G(x)$ as an asymptotic series.

Ordinarily, in problems encountered in science, asymptotic series are not convergent. In fact, it is ordinarily true that, for a given x, the numbers $|a_0|$, $|a_1 x^{-1}|$, $|a_2 x^{-2}|$, ..., form a sequence of which the first M members decrease monotonically and the subsequent members increase monotonically. Note that the value of M depends on the value of x. Note also that these ordinarily-true statements hold, in particular, for the series for $Ei(x)$, and that here, moreover, the remainder R_n is smaller in magnitude than the $(n+1)$th term of the series.

10.1 Problems

10.1.1 For what values of x can the asymptotic series for $Ei(x)$ provide a description which is in error by less than 10%? For that accuracy, where should the series be truncated for computational purposes? How many terms of the series must be retained if one is to describe $Ei(x)$ in $x > 1$ with no more than 20% error?

10.1.2 Find the asymptotic series for erf x (cf. Equation (11.1)) and estimate the range of x over which it can be used to describe erf x with 1% accuracy. How many terms should be used? Over what x can it be used to describe erf x to within an error of 10^{-4}? How many terms should be used?

10.1.3 Use the definition of an asymptotic series to show that

(a) there is only one asymptotic series $\sum a_n x^{-n}$ for a given function, $F(x)$;

(b) several functions may have the same asymptotic series [HINT: Study $Ei(x) + e^{-2x}$];

(c) the asymptotic series $\sum_{n=0}^{\infty} \alpha_n x^{-n}$ and $\sum_{n=0}^{\infty} \beta_n x^{-n}$ for $\alpha(x)$ and $\beta(x)$ (respectively) may be added {multiplied} term by term to give an asymptotic series for $\alpha(x) + \beta(x)$, $\{\alpha(x) \cdot \beta(x)\}$;

(d) if $F(x)$ and $F'(x)$ have asymptotic expansions, with

$$F(x) \sim \sum_{n=0}^{\infty} f_n x^{-n},$$

then

$$F'(x) \sim \sum_{n=0}^{\infty} - nf_n x^{-(n+1)};$$

(e) each coefficient in the asymptotic series for the function $H(x) \equiv 0$ is zero.

10.2 An Elementary Technique

Frequently, even though the solution of a given differential equation is not a simple combination of elementary functions, an asymptotic representation of that solution can be constructed in terms of elementary functions. Consider, for example, the equation

$$u''(x) + u(x) - \frac{\beta^2 - \frac{1}{4}}{x^2} u(x) = 0. \tag{10.7}$$

We note that, for very large x, the term $(\beta^2 - \frac{1}{4})u(x)/x^2$ is not likely to be very important compared to $u(x)$ and that a plausible approximation to Equation (10.7) is

$$u'' + u \cong 0.$$

If this is acceptable, then one plausible approximation to a solution is

$$u \cong e^{ix}.$$

The optimist could now hope that $u(x)$ might have an asymptotic representation in the form

$$u \sim e^{ix} \sum_{n=0}^{\infty} a_n x^{-n}. \tag{10.8}$$

To examine the consequences of this conjecture, we substitute Equation (10.8) into Equation (10.7) to obtain

$$-\sum_{n=0}^{\infty} a_n x^{-n} - 2i \sum_{n=1}^{\infty} na_n x^{-(n+1)} + \sum_{n=0}^{\infty} n(n+1)a_n x^{-(n+2)}$$
$$+ \sum_{n=0}^{\infty} a_n x^{-n} - (\beta^2 - \tfrac{1}{4}) \sum_{n=0}^{\infty} a_n x^{-n-2} = 0.$$

In particular (using the result of Problem 10.1.3 (e)), we have

$$-2ia_1 = (\beta^2 - \tfrac{1}{4})a_0, \qquad -2i(2a_2) = -2a_1 + (\beta^2 - \tfrac{1}{4})a_1,$$

and, arbitrarily choosing $a_0 = 1$,

$$u \sim e^{ix}\left[1 + \frac{i(\beta^2 - \tfrac{1}{4})}{2x} - \frac{(\beta^2 - \tfrac{1}{4})(\beta^2 - \tfrac{9}{4})}{8x^2} + \cdots\right]. \tag{10.9}$$

Note that we have tacitly assumed that the series of Equation (10.9) is an asymptotic series despite the fact that we have not estimated the size of the remainder R_n, and have not checked that remainder against the definition of an asymptotic series. It is a fact of life that, in many problems in which asymptotic series are of enormous value, the situation is too intricate to allow the investigator to make a useful estimate of the remainder and he frequently makes do with a series such as Equation (10.9) which has the form of an asymptotic series but which he cannot prove is an asymptotic series. The reader should realize that the literature contains many solutions of problems which nominally are asymptotic descriptions and which are incorrect and misleading. This does not mean that series which are only apparently asymptotic in character should be rejected out of hand, but it does mean that good judgment must accompany their construction and that readers should subject apparent asymptotic descriptions to a critical scrutiny.

Both for the foregoing particular example and for the following more general problems, it can be shown rigorously that the series so obtained are asymptotic. For such a proof, the reader is referred to Ince, *Ordinary Differential Equations*, p. 169, Dover Publications, or to Chapter 6 of Carrier, Krook and Pearson, *Functions of a Complex Variable, Theory and Technique*, McGraw-Hill, 1966 (reprinted, Hod Books, Ithaca, N.Y.).

10.3 Problems

10.3.1 Let

$$u''(x) + f(x)u'(x) + h(x)u(x) = 0,$$

and let $f(x)$ and $h(x)$ be such that they can be expanded in the form

$$f(x) \sim \sum_{n=0}^{\infty} f_n x^{-n}, \qquad h(x) \sim \sum_{n=0}^{\infty} h_n x^{-n}.$$

Using the transformations

$$u(x) = w(x)e^{\lambda x}, \quad \text{with} \quad \lambda^2 + \lambda f_0 + h_0 = 0,$$

and

$$w(x) = x^\sigma v(x), \quad \text{with} \quad (2\lambda + f_0)\sigma + \lambda f_1 + h_1 = 0,$$

show that we can obtain representations for $u(x)$, which are apparently asymptotic, and which have the form

$$u(x) \sim e^{\lambda x} x^\sigma \sum_{n=0}^{\infty} v_n x^{-n}.$$

Let $z = 1/x$, and $u(x) = q(z)$. Use these substitutions in the differential equation for $u(x)$ and give a criterion (in terms of the coefficients in the equation for u) which distinguishes those cases in which the equation for $q(z)$ has a regular singular point at $z = 0$, and those cases in which it has an irregular singular point. Under what particular conditions might there *not* be two solutions with asymptotic representations in powers of x?

10.3.2 Find the asymptotic representation of a second linearly independent solution of Equation (10.7). Find the asymptotic representations of two linearly independent real solutions of Equation (10.7). Are any of these series convergent?

10.3.3 Find an asymptotic description of a solution of

$$w''(x) + w'(x) + \frac{1}{x^2} w(x) = \frac{1}{x^3}.$$

Find an asymptotic description of that $w(x)$ for which $w(x_0) = A$ and $w'(x_0) = B$, where $x_0 \gg 1$. Justify the retention or omission of any of the contributions in your answer in connection with its use in evaluating $w(2x_0)$.

10.4 Another Technique

An alternative procedure may be used to construct an asymptotic representation of the solution of such equations as (the familiar)

$$u'' + xu = 0. \tag{10.10}$$

The coefficient of u does not fit the requirements of the foregoing procedure. To evolve an alternative procedure we reason as follows. We confine our attention to values of x such that $x \gg 1$ and ask: what would a solution, U, of Equation (10.10) be, if the coefficient x in Equation (10.10) were replaced by the large constant x_0? Clearly an answer is

$$U = Ae^{ix\sqrt{x_0}} \tag{10.11}$$

and, with $x_0 \gg 1$, a wavelength of this periodic function corresponds to a range of x of length $2\pi/\sqrt{x_0}$. Thus, since the coefficient of u in Equation (10.10) changes relatively little* as x ranges over a wavelength of U, Equation (10.11) *may* provide a reasonably accurate estimate of a solution of Equation (10.10) in the neighborhood of any large x_0. We can

---
* Write the coefficient as $x_0\{1 + [(x - x_0)/x_0]\}$.

explore this more thoroughly, using Equation (10.11) as a guide, by writing

$$u = e^{im(x)}$$

and substituting this representation of u into Equation (10.10) to obtain

$$-im'' + (m')^2 = x. \qquad (10.12)$$

The approximation which led to Equation (10.11) was one for which the right-hand side of Equation (10.12) was replaced by x_0. For this approximation, $m' = \sqrt{x_0}$, and Equation (10.11) is recovered. This suggests that, for large x,

$$(m')^2 \cong x$$

and we initiate the successive approximation scheme

$$m_0'(x) = \sqrt{x}$$

and, for $k > 0$,

$$m_k'(x) = \sqrt{x + im_{k-1}''(x)}.$$

The reader can verify that

$$m_1'(x) = \sqrt{x + i/(2\sqrt{x})}, \qquad (10.13)$$

$$m_2'(x) = \left\{ x + i \frac{1 - i/(4x^{3/2})}{2\sqrt{x + i/(2\sqrt{x})}} \right\}^{1/2} \sim \sqrt{x} + \frac{i}{4x} + \frac{5}{32} x^{-5/2} + \cdots,$$

so that

$$m(x) \sim \frac{2}{3} x^{3/2} + \frac{i}{4} \ln x - \frac{5}{48} x^{-3/2} + \cdots$$

and *

$$u(x) \sim x^{-1/4} \exp\left[\tfrac{2}{3} i x^{3/2} \{1 + O(x^{-3})\}\right]. \qquad (10.14)$$

Compare this result with the initial approximation in the interval $|x - x_0| \leq \beta \ll x_0$. Once Equation (10.14) has been obtained, it is evident that it would have been simpler to write

$$m'(x) = \sqrt{x} v(y) \sim \sqrt{x} \sum_{n=0}^{\infty} v_n y^{-n}, \qquad (10.15)$$

where $y = x^{3/2}$ and to substitute Equation (10.15) into Equation (10.12).

The reader should verify that he can recover Equation (10.14) in this way and that the algebraic manipulations are simpler than those above.

---

*By $O(s^v)$, we mean any $f(s)$ for which $|s^{-v} f(s)|$ is bounded for $s > s_0$.

10.5 Problems

10.5.1 Find the asymptotic representation of another linearly independent solution of Equation (10.10).

10.5.2 Let $u(x) = x^{1/2} F(\frac{2}{3} x^{3/2}) = x^{1/2} F(z)$ in Equation (10.10) and find the differential equation which F must obey. Note that this is an equation for which you have already studied the asymptotic form of the solutions. Compare the results of that study with Equation (10.14).

10.5.3 Find asymptotic representations of the solutions of

$$u'' + x^2 u = 0$$

by both of the foregoing methods (using a transformation similar to that of the foregoing problem).

10.5.4 Find asymptotic representations of the solutions of

$$u'' + \alpha e^{kx} u = 0,$$

where $k > 0$. Is this equation related to any we have already studied?

10.6 Asymptotic Expansions in a Parameter

It is frequently useful to describe a function by a representation which is asymptotic, not in the independent variable, but in a parameter of the problem. For example, consider the equation

$$u''(x, \lambda) - \lambda x u(x, \lambda) = 0, \tag{10.16}$$

where $\lambda \gg 1$.

We can take advantage of the procedure studied earlier and write

$$u = \exp\left[\int^x g(t, \lambda)\, dt\right], \tag{10.17}$$

substitute Equation (10.17) into Equation (10.16) and obtain

$$g' + g^2 = \lambda x. \tag{10.18}$$

Anticipating that, again, the g^2 term will dominate the g' term, we seek an asymptotic representation of $g(x, \lambda)$ in the form

$$g(x, \lambda) \sim \lambda^{1/2} \left[\sum_{n=0}^{\infty} g_n(x) \lambda^{-n/2} \right]. \tag{10.19}$$

Substituting Equation (10.19) into Equation (10.18), we obtain

$$\lambda[(g_0^2 - x) + \lambda^{-1/2}(2 g_0 g_1 + g_0') + \lambda^{-1}(g_1' + 2 g_0 g_2 + g_1^2) + \cdots] = 0.$$

Equating to zero the coefficient of each power of λ, we find

$$g_0'^2 = x, \qquad g_1 = \frac{-(\ln g_0)'}{2}, \qquad g_2 = -\frac{g_1^2 + g_1'}{2g_0}.$$

It follows that two linearly independent solutions of Equation (10.20) seem to have asymptotic behaviors of the form

$$u_1(x, \lambda) \sim x^{-1/4} \exp\{\pm \tfrac{2}{3}\lambda^{1/2} x^{3/2} + 0(\lambda^{-1/2} x^{-2/3})\}. \qquad (10.20)$$

The reader should note that, although these representations were derived as though the only criterion for validity were $\lambda \gg 1$, Equation (10.20) cannot be valid at $x = 0$ (the actual solutions must be bounded there, since $x = 0$ is an ordinary point of Equation (10.16)). Thus, Equation (10.20) is certainly not to be trusted in some neighborhood of $x = 0$, whose size we might estimate by asking for that value of x (for a given large λ) at which the first ignored term, $\lambda^{-1/2} g_2$, is comparable to $\lambda^{1/2} g_0$ and/or g_1. We see that $\lambda^{-1/2} g_2$ is as large as either $\lambda^{1/2} g_0$ or g_1, when $x\lambda^{1/3}$ is of order unity. Thus Equation (10.20) is a useful representation only for $x\lambda^{1/3} \gg 1$.

In fact, the reader should note that both this *lack of uniformity* and Equation (10.20) could have been deduced by letting $\lambda^{1/3} x = y$ and $u(x) = w(y)$ in Equation (10.16). The resulting equation for w has already been studied and the implications of its solutions, Equation (10.14), are identical with those of Equation (10.20).

The lack of uniform validity [i.e., the fact that Equation (10.20) is incapable of providing a truncated asymptotic representation for $u(x, \lambda)$ which gives an accurate estimate of $u(x, \lambda)$ for all x at some large value of λ] is a recurrent difficulty in the use of asymptotic expansions and we shall encounter it again.

A more general result of the foregoing heuristic technique is obtained when we study

$$f''(s, \lambda) = \lambda Q(s) f(s, \lambda) \qquad (10.21)$$

for $\lambda \gg 1$. We confine our attention to a range of s over which $Q(s) \neq 0$.

We write, again,

$$f(s) = \exp\left\{\int^s h(s', \lambda)\, ds'\right\}, \qquad (10.22)$$

substitute Equation (10.22) into Equation (10.21), and obtain

$$h'(s, \lambda) + h^2(s, \lambda) = \lambda Q(s). \qquad (10.23)$$

We seek an asymptotic description of h in the form

$$h(s, \lambda) = \lambda^{\frac{1}{2}} \left[\sum_{n=0}^{\infty} h_n(s) \lambda^{-n/2} \right] \qquad (10.24)$$

and the foregoing procedures give

$$f(s, \lambda) \sim [Q(s)]^{-1/4} \exp\left[\pm \lambda^{\frac{1}{2}} \int^s Q^{\frac{1}{2}}(s')\, ds' + 0(\lambda^{-\frac{1}{2}}) \right]. \qquad (10.25)$$

Consistent with the foregoing heuristic development, it can be shown rigorously that, for values of s over which $Q(s) \neq 0$, the function

$$f(s, \lambda) Q^{1/4}(s) \exp\left[\mp \lambda^{\frac{1}{2}} \int^s Q^{\frac{1}{2}}(s')\, ds' \right]$$

has an asymptotic series whose first term is unity. However, the remainder is bounded by an amount which depends on the range of s over which the description is to be invoked and the utility of the description depends strongly on the nature of the information to be inferred from the result. There are many subtleties in this topic, some of which are discussed in Chapter 18.*

10.7 Problems

10.7.1 Find a representation of $\phi(x, \lambda)$ which is asymptotic in λ when

$$\phi''(x, \lambda) - \lambda^2 \left[Q_0(x) + \frac{1}{\lambda} Q_1(x) + \frac{1}{\lambda^2} Q_2(x) \right] \phi(x, \lambda) = 0.$$

Confine your attention to values of x for which $Q_0(x) \neq 0, Q_1(x) \neq 0$, and $Q_2(x) \neq 0$.

10.7.2 Why, in Equation (10.24), did we not write

$$h(s, \lambda) = \lambda^{\frac{1}{2}} \sum_{n=0}^{\infty} h_n(s) \lambda^{-n}?$$

10.7.3 Use a perturbation procedure to treat the problem

$$\psi''(x, \epsilon) + (1 + \epsilon x^2)\psi(x, \epsilon) = e^{-x}(1 - \epsilon)$$

in $0 < x < 1$, with

$$\psi(0, \epsilon) = \epsilon \quad \text{and} \quad \psi'(0, \epsilon) = 1.$$

10.7.4 Find the differential equation satisfied exactly by the first term of the asymptotic representation (10.25), compare with Equation (10.21), and draw appropriate conclusions.

---

*The general method is usually termed the "WKB method." See chapter 6 of Carrier, Krook and Pearson, Ibid.

Special Functions | 11

There are several families of functions with which the scientist and engineer must become familiar. In one context or another many of them arise in conjunction with the study of ordinary differential equations. This chapter is a collection of definitions and problems involving such functions. Once the reader has solved these problems, he will have acquired a reasonable acquaintance with the more frequently encountered special functions.

11.1 The Error Function

The error function, erf x, is defined by

$$\operatorname{erf} x = \frac{2}{\sqrt{\pi}} \int_0^x e^{-t^2} dt \tag{11.1}$$

and the complementary error function, erfc x, is defined by

$$\operatorname{erfc} x = 1 - \operatorname{erf} x. \tag{11.2}$$

Clearly, erf x is an odd function of x. The reader can readily construct its power series representation and (in Problem 10.1.2) he already has constructed its asymptotic series. In particular,

$$\operatorname{erf}(0) = 0, \quad \operatorname{erf}'(0) = 2/\sqrt{\pi}, \quad \operatorname{erf}(\infty) = 1.$$

(1) Plot

$$\text{erf } x \text{ versus } x \quad \text{for } 0 < x < 4.$$

(2) Show that

$$\int_0^x \frac{e^{-u}}{\sqrt{u}} \, du = A \text{ erf } \sqrt{x}$$

and find A.

(3) Differentiate

$$I = \int_0^\infty e^{-a^2 x^2} \cos bx \, dx$$

with respect to b, and integrate the result by parts to obtain a relationship between dI/db and I. Deduce that

$$I = \frac{\sqrt{\pi}}{2a} e^{-b^2/4a^2}.$$

Integrate the resulting equation with respect to b in order to express the integral

$$\int_0^\infty e^{-a^2 x^2} \frac{\sin bx}{x} \, dx$$

in terms of an error function. What happens as $a \to 0$?

(4) A simple method for evaluating $J = \text{erf}(\infty)$ consists in writing

$$J^2 = \frac{4}{\pi} \int_0^\infty \int_0^\infty e^{-x^2} \cdot e^{-y^2} \, dx \, dy$$

and recognizing that this first quadrant integral can be evaluated by use of polar coordinates. Use a similar device for

$$\int_0^\infty \cos ax^2 \, dx = \lim_{\epsilon \to 0} [\text{real part of } \int_0^\infty e^{iax^2 - \epsilon x^2} \, dx].$$

(5) Differentiate

$$\int_0^\infty e^{-\alpha t^2} \, dt$$

with respect to α, to evaluate

$$\int_0^\infty e^{-t^2} t^n \, dt.$$

What happens if the upper limit is x?

11.2 The Gamma Function

The function
$$\Gamma(x) = \int_0^\infty e^{-t} t^{x-1}\, dt \tag{11.3}$$
is called the *gamma function*.

(1) Show that the integral (11.3) "converges" (i.e., exists) for all positive x.

(2) Prove that

  (a) $\rho \Gamma(\rho) = \Gamma(\rho + 1)$ (use integration by parts);

  (b) $\Gamma(1) = 1$;

  (c) $\Gamma(n) = (n-1)!$, where n is a positive integer.

  [We use (2c) to provide us with a natural definition of $0!$ via $0! = \Gamma(1) = 1$.]

(3) Show that

  (a) $\Gamma(\tfrac{1}{2}) = \pi^{1/2}$,

  (b) $(n+\alpha)(n+\alpha-1)\cdots(\alpha+1)\alpha = \Gamma(n+\alpha+1)/\Gamma(\alpha)$.

(4) The definition of $\Gamma(x)$ can be extended to negative values of x by use of (2a). Do this, and plot a curve of $\Gamma(x)$ versus x for $-3 < x < 3$.

(5) Write
$$\Gamma(p) = 2\int_0^\infty e^{-x^2} x^{2p-1}\, dx, \qquad \Gamma(q) = 2\int_0^\infty e^{-y^2} y^{2q-1}\, dy,$$
and use the device of Problem (4) of Section (11.1) to show that
$$\frac{\Gamma(p)\Gamma(q)}{\Gamma(p+q)} = 2\int_0^{\pi/2} \cos^{2p-1}\theta \sin^{2q-1}\theta\, d\theta = \int_0^1 (1-t)^{p-1} t^{q-1}\, dt. \tag{11.4}$$

(The integral on the right is termed the *beta function*, and is denoted by $B(p, q)$. Note that $B(p, q) = B(q, p)$.)

(6) Calculate $B(p, q)$ in two different ways to deduce the *duplication formula*
$$\Gamma(2p) = \frac{2^{2p-1}}{\sqrt{\pi}} \Gamma(p)\cdot \Gamma\!\left(p + \tfrac{1}{2}\right). \tag{11.5}$$

There is a rather interesting technique by means of which we can determine the asymptotic behavior of $\Gamma(x)$ as x becomes large. The transformation $t = x\tau$ in the equation for $\Gamma(x + 1)$ as given by (11.3) leads to

$$\Gamma(x+1) = x^{x+1} \int_0^\infty e^{-x\tau} \tau^x \, d\tau = x^{x+1} \int_0^\infty e^{-x(\tau - \ln \tau)} \, d\tau$$
$$= x^{x+1} e^{-x} \int_0^\infty e^{-xf(\tau)} \, d\tau, \quad (11.6)$$

where $f(\tau) = \tau - \ln \tau - 1$. A plot of $f(\tau)$ versus τ shows that $f(\tau)$ is always positive, except at the point $\tau = 1$, where it vanishes. Thus, as x becomes large and positive, the contribution to the integral from any τ-interval not near $\tau = 1$ becomes exponentially small compared to that arising from a τ-interval near $\tau = 1$. We therefore anticipate that only the behavior of $f(\tau)$ near $\tau = 1$ will be important, and since

$$f(\tau) = 0 + \tfrac{1}{2}(\tau - 1)^2 + \cdots,$$

we have, as $x \to \infty$,

$$\Gamma(x+1) \sim x^{x+1} e^{-x} \int_0^\infty e^{-(x/2)(\tau - 1)^2} \, d\tau.$$

For essentially the same reason, the lower limit of integration can be replaced by $-\infty$; the change in variables $\tau = 1 + \xi$ then leads to the final result (*Stirling's formula*),

$$\Gamma(x+1) \sim x^{x+1} e^{-x} \int_{-\infty}^\infty e^{-x\xi^2/2} \, d\xi \sim \sqrt{2\pi x}\, x^x e^{-x}.$$

The method we have used is a special case of the powerful and elegant *saddlepoint method*.*

(7) By use of the *exact* change in variables

$$\tau - \ln \tau = u^2$$

can you obtain additional terms in the asymptotic expansion for $\Gamma(x+1)$?

(8) Show that, for an appropriate function $f(t)$, we have

$$\int_a^b e^{-xt} f(t) \, dt \sim e^{-xa}[f(a)/x + f'(a)/x^2 + f''(a)/x^3 + \cdots]$$

for large positive values of x. What replaces this formula if

$$f(t) = (t - a)^k g(t),$$

with $g(t)$ well-behaved, and with $-1 < k < 1$?

---

*See Chapter 6 of Carrier, Krook and Pearson, Ibid.

11.3 Bessel Functions of Integral Order

Bessel functions can be defined in a variety of ways. One particularly useful definition of $J_n(x)$, *the Bessel function of order n*, is (for integral values of n only)

$$J_n(x) = \frac{1}{2\pi} \int_0^{2\pi} \cos[x \sin\theta - n\theta] \, d\theta = \frac{1}{2\pi} \int_0^{2\pi} e^{i(x\sin\theta - n\theta)} \, d\theta. \quad (11.7)$$

Show that

(1) $y(x) = J_n(x)$ is a solution of *Bessel's equation of order n*, that is,

$$(xy')' + xy - \frac{n^2}{x} y = 0. \quad (11.8)$$

(2) $y(x) = J_n(\alpha x)$ is a solution of

$$(xy')' + \alpha^2 xy - \frac{n^2}{x} y = 0. \quad (11.9)$$

(3) $x^{1/2} J_n(x)$ is a solution of Equation (10.7), that is, with $u = x^{1/2} J_n(x)$,

$$u'' + u - (n^2 - \tfrac{1}{4}) u/x^2 = 0.$$

(4) $J_n(x) = \sum_{r=0}^{\infty} \frac{(-1)^r (x/2)^{n+2r}}{r!(n+r)!}.$

[HINT: Observe from Equation (11.7) that the coefficient of x^p involves $\int_0^{2\pi} e^{-in\theta} \sin^p\theta \, d\theta$. Write $\sin^p\theta$ as $[(1/2i)(e^{i\theta} - e^{-i\theta})]^p$, and use the binomial theorem, noting that at most one term is of interest.]

(5) $\quad J_0'(x) = -J_1(x), \quad [x^n J_n(x)]' = x^n J_{n-1}(x),$

$$J_n'(x) = \frac{n}{x} J_n(x) - J_{n+1}(x).$$

(6)
$$J_n(x) \sim \left(\frac{2}{\pi x}\right)^{1/2} \left\{ U \cos\left(x - \frac{n\pi}{2} - \frac{\pi}{4}\right) + V \sin\left(x - \frac{n\pi}{2} - \frac{\pi}{4}\right) \right\}$$

$$(11.10)$$

for $x \gg 1$, where

$$U = 1 - \frac{(1 - 4n^2)(9 - 4n^2)}{2!(8x)^2} + \cdots,$$

§11.3]   **Bessel Functions of Integral Order**   101

$$V = \frac{1 - 4n^2}{8x} + \frac{(1 - 4n^2)(9 - 4n^2)(25 - 4n^2)}{3!(8x)^3} + \cdots.$$

[HINT: Treat $J_0(x)$ first, using Equation (11.7) with the change of variable $\sin \theta = 1 - \beta^2/x$ and the modified range of integration $0 < \theta < \pi/2$. Obtain the formula

$$\sqrt{\frac{x}{2}} J_0(x) = \int_0^\infty \cos x \cos \beta^2 \, d\beta + \int_0^\infty \sin x \sin \beta^2 \, d\beta + R(x)$$

and show (using integrations by parts, etc.) that $R(x) \to 0$ as $x \to \infty$.

To treat $J_n(x)$, use (5).]

(7)
$$J_0(x_j) = 0 \quad \text{for} \quad x_j \cong 2.405,\ 5.520,\ 8.654,\ 11.79,\ \ldots.$$

(8)
$$J_1(y_j) = 0 \quad \text{for} \quad y_j \cong 3.832,\ 7.016,\ 10.17,\ 13.32,\ \ldots.$$

(9)
$$\exp\left[\frac{x}{2}\left(t - \frac{1}{t}\right)\right] = \lim_{N \to \infty} \left[\sum_{n=0}^{N} J_n(x) t^n + \sum_{n=-1}^{-N} (-1)^n J_n(x) t^{-n}\right].$$

The function $e^{(x/2)[t-(1/t)]}$ is the *generating function* for the Bessel functions. Sketch graphs of $J_n(x)$ for n = 0, 1, 2, $\cdots$ and for $|x| < 15$.

(10) A second solution of Bessel's equation, for the case in which n is integral, is given by the *Bessel function of the second kind* which is

$$Y_n(x) = \frac{1}{\pi} \int_0^\pi \sin(x \sin \theta - n\theta)\, d\theta$$
$$- \frac{1}{\pi} \int_0^\infty [e^{n\phi} + (-1)^n e^{-n\phi}] e^{-x \sinh \phi}\, d\phi. \quad (11.11)$$

(11) For large x,

$$Y_n(x) \sim \sqrt{\frac{2}{\pi x}} \left[\sin\left(x - \frac{n\pi}{2} - \frac{\pi}{4}\right)\right.$$
$$\left. + \frac{n^2 - \frac{1}{4}}{2x} \cos\left(x - \frac{n\pi}{2} - \frac{\pi}{4}\right) + O\left(\frac{1}{x^2}\right)\right]. \quad (11.12)$$

Consequently, if we define the *Hankel functions* $H_n^{(1)}(x)$ and $H_n^{(2)}(x)$ by

$$H_n^{(1)}(x) = J_n(x) + iY_n(x), \quad H_n^{(2)}(x) = J_n(x) - iY_n(x), \quad (11.13)$$

their asymptotic behavior, for large x, will be given by

$$H_n^{(1)}(x) \sim \sqrt{2/\pi x} \exp i[x - \tfrac{1}{2}n\pi - \tfrac{1}{4}\pi],$$
$$H_n^{(2)}(x) \sim \sqrt{2/\pi x} \exp \{-i[x - \tfrac{1}{2}n\pi - \tfrac{1}{4}\pi]\}. \quad (11.14)$$

(12) $Y_0(x)$ becomes logarithmically infinite as $x \to 0$; $Y_n(x)$, for $n > 0$, becomes algebraically infinite as $x \to 0$.
[HINT: Since Y_n and J_n are solutions of Equation (11.8), Y_n can be expressed in terms of J_n by the Wronskian technique.]

(13) The equation

$$x^2 y'' + xy' - (x^2 + n^2)y = 0$$

with n integral has the pair of solutions

$$I_n(x) = e^{-in\pi/2} J_n(ix), \quad K_n(x) = \tfrac{1}{2}\pi i e^{in\pi/2} H_n^{(1)}(ix). \quad (11.15)$$

Moreover, these modified *Bessel functions* are real for real $x > 0$. Assuming that the previous asymptotic behavior formulas are valid for complex values of x*, show also that $K_n(x)$ vanishes exponentially as $x \to \infty$. Plot $I_n(x)$ and $K_n(x)$ for n = 0, 1.

(14) The solutions $u_j(x)$, λ_j of the eigenvalue problem

$$(xu')' + \lambda xu - \frac{n^2}{x} u = 0 \quad \text{in} \quad 0 < x < 1,$$

with $|u(0)| < \infty$, $u(1) = 0$ are

$$u_j = J_n(\alpha_j x), \quad \lambda_j = \alpha_j^2,$$

where the α_j are the roots of $J_n(\alpha) = 0$.

(15) The solutions $w_j(x)$, σ_j of the eigenvalue problem

$$(xw')' + \sigma xw - \frac{n^2}{x} w = 0 \quad \text{in} \quad 0 < a < x < b$$

with $w(a) = w(b) = 0$ are

$$w_j(x) = Y_n(a\sqrt{\sigma_j})J_n(x\sqrt{\sigma_j}) - J_n(a\sqrt{\sigma_j})Y_n(x\sqrt{\sigma_j}),$$

---

* Such an assumption is, in general, valid only for a restricted range of values of the argument of the complex number x. See Chapter 6 of Carrier, Krook, Pearson, ibid.

where the σ_j are such that this expression vanishes at $x = b$. Using any set of tables, find some of the σ_j and plot the corresponding w_j.

11.4 Bessel Functions of Nonintegral Order

Define the *Bessel function of the first kind, of order ν*, by

$$J_\nu(x) = \frac{1}{\pi}\int_0^\pi \cos(x\sin\theta - \nu\theta)\,d\theta - \frac{\sin\nu\pi}{\pi}\int_0^\infty e^{-(x\sinh\phi + \nu\phi)}\,d\phi, \quad (11.16)$$

where $x > 0$ and where ν is not necessarily integral. If $\nu = n$, with n integral, Equation (11.16) is consistent with Equation (11.7).

(1) Show that $J_\nu(x)$, and also $J_{-\nu}(x)$, satisfy Bessel's equation

$$(xy')' + \left(x - \frac{\nu^2}{x}\right)y = 0. \quad (11.17)$$

(2) Show that, for large positive x,

$$J_\nu(x) \sim \sqrt{\frac{2}{\pi x}}\left[\cos\left(x - \frac{\nu\pi}{2} - \frac{\pi}{4}\right) - \frac{\nu^2 - \frac{1}{4}}{2x}\sin\left(x - \frac{\nu\pi}{2} - \frac{\pi}{4}\right)\right].$$

$$(11.18)$$

(3) Show that $J_\nu(x)$ and $J_{-\nu}(x)$ form a pair of independent solutions of Equation (11.16) if ν is non integral, but that they are not independent if ν is integral. Show also that, as $x \to 0$, $J_\nu(x) \sim$ const·x^ν. [HINT: Each of J_ν and $J_{-\nu}$ must be expressible as a linear combination of two independent series solutions obtained by the methods of Chapter 4, and since $\nu > 0$, only $J_{-\nu}$ can become infinite as $x \to 0$, etc. Thus, you can avoid the labor of a series expansion of Equation (11.16).]

(4) Define the *Bessel function of the second kind, of order ν*, by

$$Y_\nu(x) = \frac{J_\nu(x)\cos\nu\pi - J_{-\nu}(x)}{\sin\nu\pi}. \quad (11.19)$$

Show that, if $\nu \to n$, with n integral, the limit of this expression approaches the right-hand side of Equation (11.11). Thus, J_ν and Y_ν are always a pair of linearly independent solutions of Bessel's equation; only if ν is nonintegral are J_ν and $J_{-\nu}$ independent solutions.

(5) Using the methods of Chapter 4 where appropriate, show that

$$J_\nu(x) = \sum_{r=0}^{\infty} \frac{(-1)^r (x/2)^{\nu+2r}}{r!\,\Gamma(\nu + r + 1)}.\qquad(11.20)$$

(6) Defining the general Hankel functions by

$$H^{(1)}_\nu(x) = J_\nu(x) + iY_\nu(x), \qquad H^{(2)}_\nu(x) = J_\nu(x) - iY_\nu(x),\qquad(11.21)$$

show that $H^{(1)}_\nu(x)$ and $H^{(2)}_\nu(x)$ are always a linearly independent pair of solutions of Bessel's equation, and determine their asymptotic behavior.

(7) Show that each of the following equations has solutions of the form $x^p Z_\nu(ax^q)$, where a, p, q, ν are constants to be determined, and where Z_ν denotes any linear combination of J_ν and Y_ν:

$$w'' - \frac{2\sigma - 1}{x} w' + w = 0,\qquad(11.22)$$

$$w'' + xw = 0,\qquad(11.23)$$

$$w'' + \alpha x^m w = 0,\qquad(11.24)$$

$$xw'' + (1 - \alpha)w' + \tfrac{1}{4}w = 0.\qquad(11.25)$$

(8) Solve

$$y'' - (a - be^{kx})y = 0\qquad(11.26)$$

in terms of Bessel functions.

(9) Find $J_{m+\frac{1}{2}}(x)$, in terms of elementary functions, for $m = -2, -1, 0, 1, 2$.

11.5 The Airy Functions

The *Airy functions* $Ai(x)$ and $Bi(x)$ are solutions of the equation

$$u'' - xu = 0.\qquad(11.27)$$

One set of definitions is

$$Ai(x) = \pi^{-1} \int_0^\infty \cos(xt + t^3/3)\, dt,\qquad(11.28)$$

$$Bi(x) = \pi^{-1} \int_0^\infty [e^{xt - t^3/3} + \sin(xt + t^3/3)]\, dt.\qquad(11.29)$$

(1) Show that $Ai(x)$ and $Bi(x)$ do satisfy Equation (11.27).

(2) Show that, for $x < 0$ (with $z = -x$),

$$Ai(x) = \tfrac{1}{3}\sqrt{z}[J_{1/3}(\tfrac{2}{3}z^{3/2}) + J_{-1/3}(\tfrac{2}{3}z^{3/2})],$$

$$Bi(x) = \sqrt{z/3}\,[J_{-1/3}(\tfrac{2}{3}z^{3/2}) + J_{1/3}(\tfrac{2}{3}z^{3/2})]\,.$$

[HINT: Show that these functions do satisfy (11.28) and, using the fact that $\int_0^\infty \cos t^3 dt = \Gamma(\tfrac{1}{3})/2\sqrt{3}$ and $\int_0^\infty t\cos t^3 dt = \Gamma(\tfrac{2}{3})/6$, show that the two sets of definitions of Ai and Bi agree in slope and in value at the origin.]

(3) Extending the definitions of Equations (11.15) to include nonintegral values of n, show that, for $x > 0$,

$$Ai(x) = \pi^{-1}(x/3)^{1/2} K_{1/3}(\tfrac{2}{3}x^{3/2})$$

and that, for $x \gg 1$,

$$Ai(x) \sim \frac{1}{2\sqrt{\pi}} x^{-1/4} \exp\left[-\tfrac{2}{3}x^{3/2}\right].$$

Show also that (again with $z = -x$), for $-x \gg 1$,

$$Ai(x) \sim \sqrt{\tfrac{1}{\pi}}\, z^{-1/4} \sin\left(\tfrac{2}{3}z^{3/2} + \tfrac{\pi}{4}\right).$$

(4) Plot $Ai(x)$ vs x for $-10 < x < 5$.

11.6 The Legendre Polynomials

The equation

$$[(1 - x^2)u']' + \lambda u = 0, \qquad (11.30)$$

called *Legendre's equation*, is another frequently encountered equation. The solutions which are most often discussed and invoked are those for which $u(x, \lambda)$ exists and is differentiable both at $x = 1$ and at $x = -1$; the range of interest is $-1 < x < 1$. We have already seen in Problem (4.3.4) that the power-series description of the solution

$$u(x) = \sum_{j=0}^{\infty} a_j x^j$$

terminates after n terms if and only if $\lambda = n(n + 1)$, where n is an integer. It can be shown (again using complex variable theory) that *only* for such values of λ is a solution of Equation (11.30) bounded both at $x = 1$ and

$x = -1$. The solutions, $P_n(x)$, corresponding to $\lambda = n(n+1)$, are called the *Legendre Polynomials*. They are defined in the first problem below.

(1) Show that

$$P_n(x) = \frac{1}{2^n n!} \frac{d^n}{dx^n} [(x^2 - 1)^n]$$

are solutions of Equation (11.27) and that the functions $(n + \frac{1}{2})^{1/2} P_n(x)$ are orthonormal in $-1 < x < 1$ with unit weighting function. Plot $P_0(x)$, $P_1(x)$, $P_2(x)$, $P_3(x)$, versus x.

(2) Show that

$$\frac{1}{(1 - 2tx + t^2)^{1/2}} = \sum_{n=0}^{\infty} t^n P_n(x) \qquad \text{(another generating function)}.$$

(3) Show that a solution of $y''(\theta) + \cot\theta\, y'(\theta) + \lambda y(\theta) = 0$, when $\lambda = n(n+1)$, is $P_n(\cos\theta)$.

(4) Show that $(n+1)P_{n+1}(x) = (2n+1)xP_n(x) - nP_{n-1}(x)$.

(5) Show that all zeros of $P_n(x)$ are real and lie in the interval $(-1, 1)$. [HINT: Use (1) and the fact that a continuously differentiable function which vanishes at two points must possess an intervening point of zero slope.]

An extremely useful description (numerical and analytical) of the properties of many of the special functions can be found in the *Handbook of Mathematical Functions*, edited by M. Abramowitz and I. A. Stegun, National Bureau of Standards, 1964 (Dover and Wiley reprints available).

The Laplace Transform | 12

Let $f(x)$ be a function defined for $x > 0$. Its *Laplace transform*, $F(s)$, is defined by

$$F(s) = \int_0^\infty e^{-sx} f(x)\, dx. \tag{12.1}$$

As with any integral involving an infinite limit, the right-hand side of Equation (12.1) is interpreted as

$$\lim_{b \to \infty} \int_0^b e^{-sx} f(x)\, dx,$$

and we restrict our attention to those functions $f(x)$ and those values of s for which this limit exists. Similarly, if $f(x)$ has an (integrable) singularity at $x = 0$, we use the interpretation

$$\int_0^\infty e^{-sx} f(x)\, dx = \lim_{\substack{b \to \infty \\ a \to 0}} \int_a^b e^{-sx} f(x)\, dx.$$

12.1 Problems

12.1.1 Derive the following transform pairs, and verify the stated restriction on s.

$f(x)$	$F(s)$
1	$\dfrac{1}{s}, \quad s>0$
$x^n, \quad n$ integral	$\dfrac{n!}{s^{n+1}}, \quad s>0$
$x^\alpha, \quad \alpha > -1$	$\dfrac{\Gamma(\alpha+1)}{s^{\alpha+1}}, \quad s>0$
e^{ax}	$\dfrac{1}{s-a}, \quad s>a$
$\cos \omega x$	$\dfrac{s}{s^2+\omega^2}, \quad s>0$
$\sin \omega x$	$\dfrac{\omega}{s^2+\omega^2}, \quad s>0$
$x^\beta e^{ax}, \quad \beta>-1$	$\dfrac{\Gamma(\beta+1)}{(s-a)^{\beta+1}}, \quad s>a$
$\dfrac{1}{x+a}, \quad a>0$	$e^{as}\text{Ei}(as), \quad s>0$
$e^{-ax^2}, \quad a>0$	$\dfrac{\sqrt{\pi}}{2}\dfrac{e^{s^2/4a}}{\sqrt{a}}\text{erfc}\left(\dfrac{s}{2\sqrt{a}}\right)$
$\dfrac{\sin at}{t}$	$\arctan\left(\dfrac{a}{s}\right), \quad s>0$
$\ln x$	$-\dfrac{1}{s}(C+\ln s), \quad s>0,$ where $C = -\displaystyle\int_0^\infty e^{-t}\ln t\,dt$ $= .577\cdots$ (Euler's constant)

12.1.2 Let

$$I = \int_0^\infty \exp\left[-t^2 - \frac{a^2}{t^2}\right] dt \quad \text{with} \quad a > 0.$$

Obtain an alternative expression for I by use of the change in variable

$t = a/\tau$, and hence show that $dI/da = -2I$. Deduce that $I = (\sqrt{\pi}/2)e^{-2a}$, and use this result to derive the transform pair (for $b > 0$)

$$f(x) = \frac{1}{\sqrt{x}} e^{-b/x}, \qquad F(s) = \sqrt{\frac{\pi}{s}} e^{-2\sqrt{bs}}, \quad s > 0.$$

12.1.3 Let $f(x)$ have the transform $F(s)$. Define

$$g(x) = \begin{cases} 0 & \text{for } 0 < x < a, \\ f(x-a) & \text{for } x > a \end{cases}$$

and let $G(s)$ be the transform of $g(x)$. Show that $G(s) = e^{-as}F(s)$.

12.1.4 Let $f(x)$ be periodic, with period L (so that $f(x+L) = f(x)$). Define

$$g(x) = \begin{cases} f(x) & \text{for } 0 < x < L, \\ 0 & \text{for } x > L \end{cases}$$

and let $G(s)$ be the transform of $g(x)$. Show that

$$F(s) = \frac{G(s)}{1 - e^{-sL}} \qquad \text{for } s > 0.$$

Use this result to show that the transform of $f(x) = |\sin \omega t|$ is

$$F(s) = \frac{\omega}{s^2 + \omega^2} \coth\left(\frac{\pi s}{2\omega}\right).$$

12.1.5 Show that two different continuous functions, $f(x)$ and $g(x)$, cannot have the same transform $F(s)$ for $s \geq a$ by proceeding as follows. Suppose that they do have the same transform; then defining $h(x) = f(x) - g(x)$, it follows that

$$\int_0^\infty e^{-sx}h(x)\,dx = 0 \qquad \text{for } s \geq a.$$

Let $e^{-x} = t$, $h(x) = m(t)\,t^{1-a}$, to show that

$$\int_0^1 t^n m(t)\,dt = 0 \qquad \text{for } n = 0, 1, 2, \ldots.$$

Deduce that $\int_0^1 p(t)m(t)\,dt = 0$ for any polynomial $p(t)$, and, using the fact that any continuous function $m(t)$ can be approximated uniformly by a polynomial (*Weierstrass' theorem*), show that it now follows that $m(t) \equiv 0$. Thus $h(x) \equiv 0$, and $f(x) \equiv g(x)$. (Observe that we have really required only that the transforms $F(s)$ and $G(s)$ coincide at a set of equidistant values of s.)

12.1.6 Prove that, if $F(s)$ exists for $s = s_0$, it also exists for $s > s_0$. Prove also that $F'(s)$ exists for $s > s_0$. State explicitly any assumptions you make concerning $f(x)$.

12.1.7 Find $f(x)$, if $F(s) = 1/(s^2 + as + b)$, where a and b are given constants. (Note that if α and β are the roots of

$$s^2 + as + b = 0, \quad F(s) = \left(\frac{1}{s-\alpha} - \frac{1}{s-\beta}\right) \cdot \frac{1}{\alpha - \beta};$$

alternatively, write

$$\frac{1}{s^2 + as + b} = \frac{1}{(s + a/2)^2 + (b - a^2/4)},$$

and use the fact that if $G(s)$ is the transform of $g(x)$, then $G(s + a)$ is the transform of $g(x) \cdot e^{-ax}$. This second device may be particularly convenient if $b > a^2/4$, that is, if α and β are complex.)

12.2 Transform of a Derivative

The usefulness of the concept of the Laplace transform arises mainly from the fact that the transform of the derivative $f'(x)$ of a function $f(x)$ is very simply related to the transform $F(s)$ of $f(x)$. In fact, integration by parts shows that

$$\int_0^\infty e^{-sx} f'(x)\, dx = [e^{-sx} f(x)]_0^\infty + s \int_0^\infty e^{-sx} f(x)\, dx = -f(0) + sF(s). \tag{12.2}$$

Similarly, we have

$$\int_0^\infty e^{-sx} f''(x)\, dx = -f'(0) - sf(0) + s^2 F(s)$$

and, in general,

$$\int_0^\infty e^{-sx} f^{(n)}(x)\, dx = -f^{(n-1)}(0) - sf^{(n-2)}(0) - \cdots - s^{n-1} f(0) + s^n F(s). \tag{12.3}$$

In all of these cases, $f(x)$ and its derivatives are assumed to be sufficiently well-behaved functions that the transform integrals exist, say for $s > a$. We have also tacitly assumed that $e^{-sx} f(x)$, for example, $\to 0$ as $x \to \infty$, for $s > a$.

As a rudimentary example of typical Laplace transform usage, consider

$$u'(x) + au(x) = \sin \beta x \quad \text{in } 0 < x < \infty, \quad u(0) = 0. \tag{12.4}$$

We multiply each term of Equation (12.4) by e^{-sx} and integrate term by term (one refers to this process concisely as *taking the Laplace transform* of Equation (12.4)) to obtain (using Equation (12.2))

$$[-u(0) + sU(s)] + aU(s) = \beta/(s^2 + \beta^2),$$

where $U(s)$ is the transform of $u(x)$. Since $u(0) = 0$, we have

$$U(s) = \frac{\beta}{(s+a)(s^2 + \beta^2)} = \frac{\beta}{a^2 + \beta^2}\left[\frac{1}{s+a} - \frac{s-a}{s^2 + \beta^2}\right]. \quad (12.5)$$

Thus, according to the table of Problem (12.1.1), we obtain

$$u(x) = \frac{\beta}{a^2 + \beta^2}\left[e^{-ax} - \cos \beta x + \frac{a}{\beta}\sin \beta x\right]. \quad (12.6)$$

Because this is the only function (cf. Problem (12.1.5)) whose transform is given by Equation (12.5), we have obtained the unique solution of the original problem (12.4). Clearly, this same result could have been obtained without the introduction of the Laplace transform; nevertheless, this technique involves only simple algebraic manipulations and we already anticipate the fact that the use of Laplace transforms permits the replacement of certain differential equations by corresponding algebraic equations.

12.3 Problems

12.3.1 Let $u''(x) + au'(x) + bu(x) = e^{cx}$ in $0 < x < \infty$, where $u(0) = \alpha$, $u'(0) = \beta$ and where a, b, c, α, β are given constants. Show that $U(s)$, the transform of $u(x)$, satisfies the equation

$$U(s) = \frac{1}{(s-c)(s^2 + as + b)} + \frac{\alpha(s+a) + \beta}{s^2 + as + b}.$$

Expand this expression in partial fractions, and use the table of Problem (12.1.1) to determine $u(x)$. (Do not exclude the case in which the roots of $s^2 + as + b = 0$ are equal.) Which part of the above transform expression relates to the particular integral, and which to the complementary function?

12.3.2 Let $f(x)$ and $g(x)$ have the transforms $F(s)$ and $G(s)$ respectively. Show that

(a) if $g(x) = e^{ax}f(x)$, then $G(s) = F(s-a)$ (note the use of this result in finding inverse transforms; cf. Problem 12.1.7);

(b) if $g(x) = xf(x)$, then $G(s) = -\dfrac{d}{ds} F(s)$;

(c) if $g(x) = x^n f(x)$, then $G(s) = (-1)^n \dfrac{d^n}{ds^n} F(s)$;

(d) if $g(x) = f(ax)$, then $G(s) = \dfrac{1}{a} F\left(\dfrac{s}{a}\right)$;

(e) if $g(x) = xf'(x)$, then $G(s) = -\dfrac{d}{ds}[sF(s)]$;

(f) if $g(x) = [xf'(x)]'$, with $xf' \to 0$ as $x \to 0$, then $G(s) = -s\dfrac{d}{ds}[sF(s)]$.

12.3.3 Let
$$v'(x) + a_{11}v(x) + a_{12}w(x) = h_1(x), \quad w'(x) + a_{21}v(x) + a_{22}w(x) = h_2(x)$$
in $0 < x < \infty$, with $v(0) = \alpha$, $w(0) = \beta$. Denoting transforms of functions by the corresponding capital letters, what algebraic equations must $V(s)$ and $W(s)$ satisfy? Solve the special case in which $a_{11} = 1$, $a_{12} = 2$, $a_{21} = 2$, $a_{22} = -1$, $h_1(x) = 0$, $h_2(x) = e^{-x}$, and in which α, β are arbitrary constants. Can you find α, β such that $v(0) = 1$, $w(\infty) = 0$?

12.3.4 Find $U(s)$, the transform of $u(x)$, if

(a) $[xu'(x)]' + \beta^2 x u(x) = 0$ (assume $xu' \to 0$ as $x \to 0$),

(b) $u''(x) + xu'(x) + u(x) = 0$ with $u(0) = 1$, $u'(0) = 0$,

(c) $u''(x) + a^2 u(x) = \cos \beta x$ in $0 < x < L$ with $u(0) = 1$, $u(L) = 0$.

12.3.5 (a) Show that, if the transform $F(s)$ of $f(x)$ exists for $s > s_0$, then $F(s) \to 0$ as $s \to \infty$. This is a useful test which can frequently be used to show that some particular function (for example, $\sin \omega s$) cannot possibly be a Laplace transform.

(b) Let
$$f(x) = \begin{cases} 0 & \text{for } 0 < x < \alpha, \\ \dfrac{1}{\epsilon} & \text{for } \alpha < x < \alpha + \epsilon, \\ 0 & \text{for } x > \alpha + \epsilon. \end{cases}$$

Show that

$$F(s) = e^{-s\alpha}\left(\frac{1 - e^{-s\epsilon}}{s\epsilon}\right)$$

and that $F(s) \to e^{-s\alpha}$ as $\epsilon \to 0$. Does the special case corresponding to $\alpha = 0$ contradict the result of Part (a)?

12.3.6 (a) State explicitly the requirements that must be fulfilled by $f(x)$ if the following sequence of operations is to be legitimate for $s > a$:

$$F(s) = \int_0^\infty e^{-sx} f(x)\, dx = \left[\frac{e^{-sx}}{-s} f(x)\right]_0^\infty + \frac{1}{s}\int_0^\infty e^{-sx} f'(x)\, dx$$

$$= \frac{f(0)}{s} + \frac{1}{s}\int_0^\infty e^{-sx} f'(x)\, dx \Rightarrow \lim_{s \to \infty} sF(s) = f(0).$$

(b) Use the device of Part (a) and the result of Problem 12.3.4a, to show that the transform of $J_0(\beta x)$ is $1/[s^2 + \beta^2]^{1/2}$.

12.3.7 Let $f(x)$ and $g(x)$ have the transforms $F(s)$ and $G(s)$, respectively. Let $G(s) = sF(s)$. Deduce that $g(x) = f'(x)$. (Why does this result not contradict Equation (12.2)?) Again, this is a very useful device to use in inverting transforms; thus the inverse of

$$s\left(\frac{1}{s+a}\right)^2$$

can be written at once as

$$\frac{d}{dx}\left\{\text{inverse of }\left(\frac{1}{s+a}\right)^2\right\} = \frac{d}{dx}\{xe^{-ax}\} = e^{-ax} - axe^{-ax}.$$

Note also the inverse result, viz., that $F(s) = (1/s)G(s)$ will imply that $f(x) = \int_0^x g(\xi)\, d\xi$; why is there no arbitrary constant of integration here?

12.3.8 Let

$$y''(x) + a^2 y(x) = \sin x \quad \text{in} \quad 0 < x < \infty$$

and use the Laplace transform method to find *all* periodic solutions $y(x)$ for which $y(0) = 0$. For what values of a does no periodic solution exist? For what values of a is there only one periodic solution? What is the inverse transform of $(s^2 + 1)^{-2}$?

12.4 The Convolution Integral

Let $f(x)$, $g(x)$, $h(x)$ have the transforms $F(s)$, $G(s)$, $H(s)$, respectively, and let $h(x)$ be the *convolution integral* of $f(x)$ and $g(x)$; that is,

$$h(x) = \int_0^x f(x - \xi)g(\xi)\, d\xi = \int_0^x g(x - \xi)f(\xi)\, d\xi. \qquad (12.7)$$

One of the more useful identities associated with the Laplace transform technique is that

$$H(s) = F(s)G(s). \qquad (12.8)$$

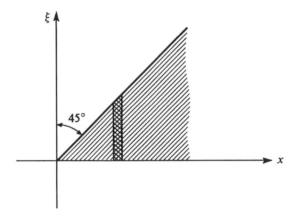

Figure 12.1 Change in order of integration.

To verify Equation (12.8), we write

$$H(s) = \int_0^\infty e^{-sx} h(x)\, dx = \int_0^\infty e^{-sx}\, dx \cdot \int_0^x f(x - \xi)g(\xi)\, d\xi$$

$$= \iint_A e^{-sx} f(x - \xi)g(\xi)\, dx\, d\xi,$$

where the area integral is over the shaded region in Figure 12.1 (this is so because we first integrate with respect to ξ between $\xi = 0$ and $\xi = x$ [the heavily shaded strip] and then with respect to x from 0 to ∞). However, the integration over the shaded area can also be accomplished by first integrating with respect to x from ξ to ∞, and then with respect to ξ from 0 to ∞ (that is, by using horizontal strips). Therefore, we have

$$H(s) = \int_0^\infty d\xi \int_\xi^\infty e^{-sx} f(x-\xi) g(\xi) \, dx$$

$$= \int_0^\infty g(\xi) e^{-s\xi} d\xi \int_\xi^\infty e^{-s(x-\xi)} f(x-\xi) \, d(x-\xi),$$

where we have used the fact that $d(x - \xi) = dx$ since ξ is fixed during the x integration. It follows that

$$H(s) = \int_0^\infty g(\xi) e^{-s\xi} d\xi \cdot \int_0^\infty e^{-s\eta} f(\eta) \, d\eta,$$

and the two integrals are now independent; hence

$$H(s) = G(s)F(s),$$

and the desired result is proved.

A typical use of this identity follows. Let

$$u''(x) + a^2 u(x) = g(x) \quad \text{in } 0 < x < \infty \tag{12.9}$$

with $u(0) = u'(0) = 0$. Taking the Laplace transform of Equation (12.9), we obtain

$$U(s) = \frac{G(s)}{s^2 + a^2},$$

and since $(s^2 + a^2)^{-1}$ is the transform of $a^{-1} \sin ax$, Equations (12.7) and (12.8) imply that $u(x)$ is given by either

$$u(x) = \frac{1}{a} \int_0^x \sin a(x - \xi) g(\xi) \, d\xi$$

or $\tag{12.10}$

$$u(x) = \frac{1}{a} \int_0^x g(x - \xi) \sin a\xi \, d\xi.$$

The simplicity of the foregoing algebra is noteworthy.

12.5 Problems

12.5.1 Use the convolution theorem to invert the transforms

(a)
$$\frac{1}{(s+1)^2},$$

(b)
$$\left(\frac{1}{s-a}\right) \frac{s}{s^2 + \omega^2},$$

(c)
$$(s-a)^{-\pi}\arctan\left(\frac{3}{s}\right),$$

(d)
$$\frac{1}{s}e^{-a\sqrt{s}}$$

(compare Problem 12.1.2).

12.5.2 (a) Let $F(s)$, $G(s)$, $H(s)$ be the transforms of $f(x)$, $g(x)$, $h(x)$, respectively; develop a formula for the inverse transform of $F(s)G(s)H(s)$.

(b) Use Laplace transforms to prove the identity
$$\int_0^x dx_1 \int_0^{x_1} dx_2 \cdots \int_0^{x_{n-1}} f(x_n)\, dx_n = \int_0^x \frac{(x-\xi)^{n-1}}{(n-1)!} f(\xi)\, d\xi.$$

12.5.3 Use Laplace transforms to obtain compactly a particular integral of
$$y'' + ay' + by = f(x),$$
where a, b are constants.

12.6 Asymptotic Behavior

We have seen in Problem (12.3.6) (and also in Problem 8 of Sec. 11.2) that the behavior of $f(x)$ for small values of x is related to the behavior of its transform, $F(s)$, for large values of s. That such a relationship should exist is very reasonable, for, as s becomes larger and larger in
$$F(s) = \int_0^\infty e^{-sx} f(x)\, dx,$$
it is clear that the integral will tend to become more and more dominated by contributions from an interval close to the origin. We now develop this relationship in general.

Let
$$f(x) = x^\nu [a_0 + a_1 x + a_2 x^2 + \cdots + a_m x^m + R_m(x)] \quad (12.11)$$
for $0 < x < A$, with $\nu > -1$ and with $|R_m(x)| < Cx^{m+1}$, where A and C are constants. We also assume that $|f(x)| < Be^{s_0 x}$ for some constants B and s_0, and for $x > A$. Then, for $s > s_0$, we have

$$F(s) = \int_0^A e^{-sx} x^\nu [a_0 + a_1 x + \cdots + R_m(x)] \, dx + \int_A^\infty e^{-sx} f(x) \, dx.$$

As $s \to \infty$, the second integral is of order $e^{-(s-s_0)A}$, so that (as we will see) its contribution for large s can be neglected in comparison with contributions from the first integral. Thus we obtain

$$F(s) \sim \int_0^A e^{-sx} x^\nu [a_0 + a_1 x + \cdots + a_m x^m + R_m(x)] \, dx.$$

We write

$$\int_0^A e^{-sx} x^\nu (a_k x^k) \, dx = \int_0^\infty e^{-sx} x^\nu a_k x^k \, dx - \int_A^\infty e^{-sx} x^\nu a_k x^k \, dx,$$

$$= \frac{a_k \Gamma(\nu + k + 1)}{s^{\nu+k+1}} + \text{order of } (e^{-As}).$$

Thus, finally, we have

$$F(s) \sim a_0 \frac{\Gamma(\nu + 1)}{s^{\nu+1}} + a_1 \frac{\Gamma(\nu + 2)}{s^{\nu+2}} + \cdots + a_m \frac{\Gamma(\nu + m + 1)}{s^{\nu+m+1}} \quad (12.12)$$

as $s \to \infty$, where the error term arising from $R_m(x)$ is of order $1/s^{\nu+m+2}$ at most. This result is referred to as *Watson's Lemma*. Why did we require $\nu > -1$?

12.7 Problems

12.7.1 Find the Laplace transform of $x^\nu J_\nu(x)$, with $\nu > 0$. (Start with the differential equation, one of whose solutions is $w = x^\nu J_\nu(x)$.) What assures you that your answer is not the transform of $Ax^\nu J_\nu(x) + Bx^\nu J_{-\nu}(x)$, with $B \neq 0$? Why is it more difficult, using the same procedure, to find the transform of $x^{-\nu} J_\nu(x)$?

12.7.2 Let

$$h(x) = \int_0^x f(t) \cdot g(x - t) \, dt,$$

where $t^{-\nu} f(t) \to A$ as $t \to 0$, $t^{-\mu} g(t) \to B$ as $t \to 0$, with $\nu > -1$, $\mu > -1$. How does $h(x)$ behave near $x = 0$?

12.7.3 (a) Given that $F(s) = \ln[1 + (1/s)]$, expand $F(s)$ in powers of $1/s$ for large s so as to find a series expression for $f(t)$. Sum the series.

(b) Solve the same problem by first deciding how the inverse transform of dF/ds is related to $f(t)$.

12.7.4 Figure 12.2 depicts an electric circuit sometimes used to approximately "differentiate" an input signal $e_i(t)$. The transform $E_o(s)$ of the output voltage $e_o(t)$ is related to the transform $E_i(s)$ of $e_i(t)$ by the circuit algebra formula *

$$E_o(s) = \frac{R}{R + (1/Cs)} E_i(s),$$

where R and C are circuit constants. What relations should R and C satisfy in order that $e_o(t) \cong e_i'(t)$? How good is this approximation?

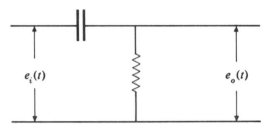

Figure 12.2

12.8 Partial Differential Equations

It would be unfortunate to omit some illustration of the much greater power that the Laplace transform can provide in treating problems which involve partial differential equations. Accordingly, we digress momentarily from our study of ordinary differential equations to consider the following problem.

Find $\phi(x, t)$ where

$$\phi_{xx} = \phi_t \quad \text{in } 0 < t < \infty, \quad -L < x < L \qquad (12.13)$$

with

$$\phi(-L, t) = \phi(L, t) = 0 \quad \text{for } t > 0 \qquad (12.14)$$

and

$$\phi(x, 0) = 1. \qquad (12.15)$$

We define the transform with respect to t of $\phi(x, t)$ by

---

*If a circuit containing passive circuit elements has neither current flowing nor charges on any capacitors at time zero, then it is a familiar fact to electrical engineers that Kirchoff's circuit law ₊pply to the transforms of currents and voltages, provided that the impedance of an inductor L, a capacitor C, and a resistor R are written Ls, $1/Cs$, and R, respectively.

$$\Phi(x, s) = \int_0^\infty e^{-st}\phi(x, t)\, dt,$$

and we take the transform of Equation (12.13) (i.e., we multiply each term by e^{-st} and integrate with respect to t from 0 to ∞). Taking account of Equation (12.15), we obtain the ordinary differential equation

$$\Phi_{xx} = -1 + s\Phi, \tag{12.16}$$

where we have assumed that the orders of integration and differentiation can be interchanged, so that

$$\int_0^\infty e^{-st}\phi_{xx}(x, t)\, dt = \left(\int_0^\infty e^{-st}\phi(x, t)\, dt\right)_{xx} = \Phi_{xx}(x, s).$$

Equation (12.16) implies that

$$\Phi = (1/s) + Ae^{\sqrt{s}\, x} + Be^{-\sqrt{s}\, x}, \tag{12.17}$$

where A and B are constants of integration. Since s is a parameter of Equation (12.16), and since any function of s is a constant insofar as differentiation with respect to x is concerned, A and B may actually be functions of s. In fact, the boundary conditions $\phi(-L, t) = \phi(L, t) = 0$ imply that $\Phi(-L, s) = \Phi(L, s) = 0$, so that

$$A = B = -\frac{1}{2s\cosh\sqrt{s}\, L}$$

and

$$\Phi = \frac{1}{s}\left[1 - \frac{\cosh\sqrt{s}\, x}{\cosh\sqrt{s}\, L}\right]. \tag{12.18}$$

We have thus obtained the transform of $\phi(x, t)$; it may be inverted by means of tables, or by the method of Problem (12.9.1), to give $\phi(x, t)$ itself.

12.9 Problems

12.9.1 For large s, the term $\cosh\sqrt{s}\, x / \cosh\sqrt{s}\, L$ in Equation (12.18) can be written

$$\frac{\cosh\sqrt{s}\, x}{\cosh\sqrt{s}\, L} = \frac{e^{\sqrt{s}\, x} + e^{-\sqrt{s}\, x}}{e^{\sqrt{s}\, L}(1 + e^{-2\sqrt{s}\, L})}$$
$$= e^{-\sqrt{s}\, L}(e^{\sqrt{s}\, x} + e^{-\sqrt{s}\, x})(1 - e^{-2\sqrt{s}\, L} + e^{-4\sqrt{s}\, L} + \cdots)$$

and the resulting transform expression can then be inverted term by term, by use of the result of Problem 12.1.2. Carry out this process so as to obtain an expression for $\phi(x, t)$ valid at least for small t (the result is actually valid for all t).

12.9.2 Obtain the transform of $\phi(x, t)$ for the problem of Equation (12.13) and Equation (12.14), with Equation (12.15) modified to read

$$\phi(x, 0) = f(x).$$

What can you deduce concerning the behavior of $\phi(x, t)$ for small t?

12.9.3 Solve the problem

$$\phi_{xx} = \phi_t \quad \text{in} \quad 0 < t < \infty, \quad -\infty < x < \infty;$$

$$\phi(x, t) \quad \text{bounded for all} \quad x, t;$$

$$\phi(x, 0) = f(x).$$

[HINT: Write

$$\Phi = e^{\sqrt{s}\,x}\left[A - \frac{1}{2\sqrt{s}}\int_0^x f(\xi)e^{-\sqrt{s}\,\xi}\,d\xi\right]$$

$$+ e^{-\sqrt{s}\,x}\left[B + \frac{1}{2\sqrt{s}}\int_0^x f(\xi)e^{\sqrt{s}\,\xi}\,d\xi\right]$$

and decide what A and B must be in order that Φ remain bounded as $x \to \pm\infty$.] Obtain the final result

$$\phi(x, t) = \frac{1}{2\sqrt{\pi t}}\int_{-\infty}^\infty f(\xi)\exp\left[-\frac{(x-\xi)^2}{4t}\right]d\xi. \qquad (12.19)$$

How does the function $(1/2\sqrt{\pi t})e^{-(x-\xi)^2/4t}$ behave as $t \to 0$? Compute the area under this function, for any t.

The power of the Laplace transform and of other integral transforms is greatly augmented by a thorough understanding of the theory of functions of a complex variable. Further reading along the lines of *Functions of a Complex Variable: Theory and Technique*, by Carrier, Krook and Pearson, ibid.

Rudiments of the Variational Calculus | 13

Another form into which many boundary value problems can be cast is readily approached in the following way. We ask: can we choose among all functions, $y(x)$, for which $y''(x)$ exists and is continuous at each point in the interval $a \leq x \leq b$, that one which minimizes the integral

$$I = \int_a^b [(y'(x))^2 + y^2(x) - xy(x)]\, dx? \tag{13.1}$$

To answer this, we denote the particular function we seek by $w(x)$, and we consider

$$y(x) = w(x) + \epsilon\eta(x), \tag{13.2}$$

where $\eta(x)$ is an arbitrary (sufficiently differentiable) function. For any given function $\eta(x)$, a whole family of functions, $y(x)$, is defined according to the value of ϵ. Thus, for a given η, I is a function of ϵ and we require $I(\epsilon)$ to attain a minimum at $\epsilon = 0$.

Since $I(\epsilon)$ is a quadratic function of ϵ, the condition that $I(0) < I(\epsilon)$ implies that

$$\left(\frac{\partial I}{\partial \epsilon}\right)_{\epsilon=0} = 0.$$

We write, for any particular $\eta(x)$,

$$I(\epsilon) = \int_a^b [(w' + \epsilon\eta')^2 + (w + \epsilon\eta)^2 - x(w + \epsilon\eta)]\, dx, \tag{13.3}$$

and we differentiate $I(\epsilon)$ to find

$$\left(\frac{\partial I}{\partial \epsilon}\right)_{\epsilon=0} = \int_a^b [2\eta'w' + 2\eta w - \eta x]\, dx = 0. \tag{13.4}$$

Integration by parts yields

$$-\int_a^b \eta[2w'' - 2w + x]\, dx + 2\eta(b)w'(b) - 2\eta(a)w'(a) = 0. \tag{13.5}$$

Suppose now that $2w'' - 2w + x \neq 0$ at some point x_0 in (a, b). Then, since w and w'' are continuous, $2w'' - 2w + x$ is one-signed in some finite region, $a < A < x < B < b$. Since Equation (13.5) must hold for any $\eta(x)$ for which $\eta''(x)$ exists and is continuous in (a, b), we choose as a particular $\eta(x)$ the function

$$\eta(x) = \begin{cases} [(B - x)(x - A)]^3 & \text{in } A < x < B, \\ 0 & \text{elsewhere.} \end{cases} \tag{13.6}$$

Clearly, Equation (13.5) is violated by this choice because the integrand is one-signed at every point of a region of finite extent and $\eta(a) = \eta(b) = 0$. This contradiction stems from the hypothesis that $2w'' - 2w + x \neq 0$ for at least one point x_0; therefore, we conclude that

$$2w'' - 2w + x = 0 \quad \text{in } a < x < b. \tag{13.7}$$

Thus the integral term in Equation (13.5) vanishes. In order to satisfy Equation (13.5) for *all* admissible functions, η, we must still require that

$$2\eta(b)w'(b) - 2\eta(a)w'(a) = 0 \tag{13.8}$$

and when we use such choices for $\eta(x)$ as $\eta_1(x) = x - a$ and $\eta_2(x) = x - b$, this is accomplished only if

$$w'(a) = w'(b) = 0. \tag{13.9}$$

It follows that any function $w(x)$ which minimizes I, as required in the initial statement of the problem, must also be a solution of the boundary-value problem

$$2w''(x) - 2w(x) + x = 0 \quad \text{in } a < x < b \tag{13.10}$$

with

$$w'(a) = w'(b) = 0. \tag{13.11}$$

The differential Equation (13.10) is called the *Euler equation* of this variational problem and Equations (13.11) are the *natural boundary conditions*.

It does not follow from the foregoing that any function $w(x)$ which is a solution of this boundary value problem must *minimize I*. It does follow from Equation (13.4) that for such a $w(x)$, $I(\epsilon)$ has zero slope at $\epsilon = 0$; this might however indicate a maximum rather than a minimum. The appropriate terminology is: any $w(x)$ which satisfies Equation (13.10) and Equation (13.11) *renders I stationary*. Thus, in order that $w(x)$ should satisfy the boundary-value problem of Equations (13.10) and (13.11) it is both necessary and sufficient that $w(x)$ should render I stationary.

For the foregoing problem, the reader can verify that

$$w(x) = \frac{x}{2} - \frac{\sinh [x - (b + a)/2]}{2 \cosh [(b - a)/2]}. \tag{13.12}$$

He can also verify (using Equation (13.10) and Equation (13.1)) directly that

$$I(w + h) \geq I(w)$$

for any twice differentiable function $h(x)$. Thus, in this particular problem, $w(x)$ *does* minimize I.*

There is a variation on the foregoing problem which is also of great importance. We ask: among all functions $y(x)$ for which $y''(x)$ exists and is continuous in $a < x < b$ and for which $y(a) = y(b) = 0$, can we find that one which minimizes I where

$$I = \int_a^b [(y'(x))^2 + y^2(x) - xy(x)]\, dx? \tag{13.13}$$

Note that this question differs from that we studied above only in that the class of *admissible functions*, $y(x)$, is more restricted in this problem.

We again denote by $w(x)$ the admissible function which does minimize I (so that $w(a) = w(b) = 0$), and we define, again,

$$y(x) = w(x) + \epsilon\eta(x), \tag{13.14}$$

where $\eta(x)$ is any admissible function; i.e., any function satisfying $\eta(a) = \eta(b) = 0$. We substitute Equation (13.14) into Equation (13.13) and differentiate to obtain

---

*Ordinarily, in subtler variational problems, it can be very difficult to establish that the variational integral is actually minimized by the solution of the corresponding boundary-value problem. In many applications, however, the important feature of the variational process is that a differential equation problem becomes equivalent to the problem of rendering a certain integral stationary; whether the stationary point corresponds to a maximum, a minimum, or even an inflection point is of secondary importance.

$$\left(\frac{dI}{d\epsilon}\right)_{\epsilon=0} = \int_a^b [2\eta'w' + 2\eta w - \eta x]dx = 0. \tag{13.15}$$

Integration by parts again leads to

$$-\int_a^b \eta[2w'' - 2w + x]\,dx + 2\eta(b)w'(b) - 2\eta(a)w'(a) = 0. \tag{13.16}$$

However, since $\eta(a) = \eta(b) = 0$, it follows that each of the last two terms in Equation (13.16) is zero; thus, we require only that

$$\int_a^b \eta(x)[2w'' - 2w + x]\,dx = 0. \tag{13.17}$$

The reasoning we used before leads immediately to the conclusion

$$2w''(x) - 2w(x) + x = 0 \tag{13.18}$$

in $a < x < b$, where, as we required earlier,

$$w(a) = w(b) = 0. \tag{13.19}$$

Thus, the alternative questions we have posed have answers which are, respectively, the solutions of two different boundary value problems.*

13.1 Problems

13.1.1 Complete the solution of the problem posed by Equations (13.18) and (13.19).

13.1.2 Let

$$I = \int_a^b [p(x)(u'(x))^2 - (q(x) + \lambda r(x))u^2(x) + 2f(x)u(x)]\,dx, \tag{13.20}$$

where $p(x)$, $q(x)$, $r(x)$, and $f(x)$ are sufficiently nonpathological that no operations you are tempted to try are invalid. Show that the answer to the question "Among all functions $u(x)$ for which $u''(x)$ is continuous in $a < x < b$, which one or ones render I stationary?" is also the answer to "Among all functions $u(x)$ for which $u''(x)$ is continuous in $a < x < b$, which one or ones satisfy

$$(p(x)u'(x))' + (q(x) + \lambda r(x))u(x) = f(x) \tag{13.21}$$

with $u'(a) = u'(b) = 0$?

13.1.3 Construct a variational problem whose solution is that function $w(x)$ such that

---
*Although we specified $y(a) = y(b) = 0$ in the problem of Equation (13.13), other choices could be made; Equation (13.19) would then be modified appropriately. If only one of $y(a)$ or $y(b)$ were specified, a natural boundary condition would be applied at the other end point.

with
$$xw''(x) + w'(x) + xw(x) = 0 \tag{13.22}$$

$$w'(1) = 0 \quad \text{and} \quad w(2) = 1. \tag{13.23}$$

13.1.4 Let
$$J = \int_a^b F[y'(x), y(x), x]\, dx.$$

Let $f(x)$ be that $y(x)$ which, among those $y(x)$ with $y''(x)$ continuous in $a < x < b$, renders J stationary. Find the Euler equation and natural boundary conditions which f must satisfy. F is a given function of y', y, and x.

13.1.5 Find the number A such that the function which minimizes $I + Au^2(b)$ must obey the natural boundary condition
$$u'(b) + au(b) = 0.$$
Here I is the integral of Problem (13.1.2).

13.1.6 What differential equation and boundary conditions must $w(x)$ satisfy if
$$I\{y(x)\} = \int_0^L \{[y''(x)]^2 + r(x)[y'(x)]^2 + q(x)[y(x)]^2\}\, dx,$$
where $w(x)$ is that function for which $w^{IV}(x)$ is continuous in $0 < x < L$ and
$$I\{w(x)\} < I\{y(x)\}$$
for all other functions $y(x)$ with continuous $y^{IV}(x)$ in $0 < x < L$? (Note that r and q need not be positive in $(0,L)$.)

13.1.7 What variational problem is equivalent to the boundary-value problem
$$f^{IV}(x) - k^2 f(x) = h(x)$$
(a) with
$$f(0) = f''(0) = f'(L) = 0 \quad \text{and} \quad f(L) = 1;$$
(b) with
$$f(0) = f(L) = f'(L) = f''(0) + Af'(0) = 0?$$

13.2 The Solution of Differential Equations

Although problems in the calculus of variations arise in many contexts, we are concerned here with the manner in which this discipline provides a technique for solving boundary-value problems of ordinary differential equations. To illustrate such a technique, which frequently is called *the direct method of the variational calculus,* we address ourselves to the problem wherein

$$\alpha u'' - u = -1 \quad \text{in} \quad |x| < 1 \tag{13.24}$$

with $u(-1) = u(1) = 0$; here we know the exact solution for comparison.

According to Problem (13.1.2), this is equivalent to asking for that function $u(x)$ which, among all * $y(x)$ satisfying $y(-1) = y(1) = 0$, renders stationary the integral

$$I = \int_{-1}^{1} [\alpha(y')^2 + y^2 - 2y] \, dx \tag{13.25}$$

We might hope that a good approximation to $u(x)$ might be found if we modify the question to read; among all functions of the form

$$y_1(x) = (a + b) - ax^2 - bx^4, \tag{13.26}$$

which one renders I stationary? This, of course, is equivalent to asking: among all numbers a and b, which pair renders the right-hand side of Equation (13.25) stationary? We note that, for any values of a and b, the function $y_1(x)$ satisfies the desired boundary conditions $y_1(-1) = y_1(1) = 0$. When we substitute Equation (13.26) into Equation (13.25), we obtain

$$I(a, b) = \alpha \left(\frac{4a^2}{3} + \frac{16ab}{5} + \frac{16b^2}{7} \right) + (a + b)^2 - \frac{2a(a + b)}{3} - \frac{2b(a + b)}{5}$$

$$+ \frac{a^2}{5} + \frac{2ab}{7} + \frac{b^2}{9} - 2(a + b) + \frac{2a}{3} + \frac{2b}{5}. \tag{13.27}$$

Here $I(a, b)$ is stationary at that (a, b) for which

$$I_a = I_b = 0.$$

(Incidentally, we could have carried out this partial differentiation process under the integral sign in Equation (13.25), before evaluating I in the form (13.27); the same result, of course, would be recovered.) The reader can verify that this leads to

$$a = \frac{7}{4} \frac{18\alpha - 1}{1 + 28\alpha + 63\alpha^2}, \quad b = \frac{21}{8} \frac{1}{1 + 28\alpha + 63\alpha^2},$$

so that the required function (call it $\phi(x)$) is

---

* We omit, henceforth, the phrase *twice differentiable*, which is implicit in all that we do here.

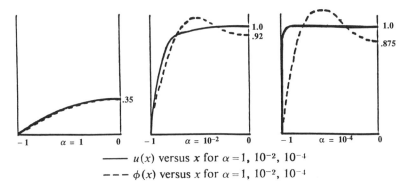

——— $u(x)$ versus x for $\alpha = 1, 10^{-2}, 10^{-4}$
- - - $\phi(x)$ versus x for $\alpha = 1, 10^{-2}, 10^{-4}$

Figure 13.1

$$\phi(x) = \frac{7[1 + 36\alpha - (36\alpha - 2)x^2 - 3x^4]}{8(1 + 28\alpha + 63\alpha^2)}. \qquad (13.28)$$

This can be compared with the exact solution of the original problem; that is,

$$u(x) = 1 - \frac{\cosh(x/\sqrt{\alpha})}{\cosh(1/\sqrt{\alpha})} \qquad (13.29)$$

or, in a form more convenient for negative α,

$$u(x) = 1 - \frac{\cos(x/\sqrt{-\alpha})}{\cos(1/\sqrt{-\alpha})}. \qquad (13.30)$$

Figures 13.1 and 13.2 characterize the adequacy of the approximation.

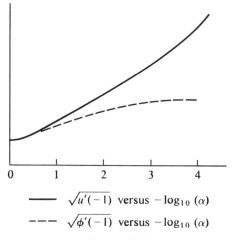

——— $\sqrt{u'(-1)}$ versus $-\log_{10}(\alpha)$
- - - $\sqrt{\phi'(-1)}$ versus $-\log_{10}(\alpha)$

Figure 13.2

For $\alpha < 0$, the adequacy of $\phi(x)$ as an approximation to $u(x)$ is easily characterized by comparing the values of α at which each is undefined. The denominator of $u(x)$ vanishes, for each integer n, at $-\alpha = [\pi(2n+1)/2]^{-2}$; that of $\phi(x)$ vanishes at $-\alpha = .405 \cong (2/\pi)^2$ and at $-\alpha = .0392 \cong .9(2/3\pi)^2$. A more detailed comparison of the implications of Equations (13.28) and (13.29) reveals that $\phi(x)$ is indeed a poor approximation to $u(x)$ for $-.039 < \alpha < 0$.

There are several important observations which can now be inferred; the reader should verify the truth of each.

(1) When $\alpha > 0$, $I\{u(x)\} < I\{y(x)\}$ for all admissible $y(x) \neq u(x)$.

(2) When $\alpha > 0$, $\phi(x)$ is an excellent approximation to $u(x)$ for any α which doesn't require $u(x)$ to be terribly steep, and is qualitatively a good approximation for all positive α.

(3) When $\alpha < 0$, $\phi(x)$ is a good approximation for some values of α, but is hopelessly inadequate for some other values.

(4) *The approximation supplied by the foregoing procedure can only be as good as the selection of the functions $y_1(x)$ to be used in the method.*

(5) If $y_n(x) = \epsilon \sin n\pi x$, then $I(u + y_n(x)) = I(u) + \epsilon^2(1 + \alpha n^2\pi^2)$. Thus, for $\alpha < 0$ with $|\alpha| < 1/\pi^2$, $I(u)$ is neither a maximum nor a minimum, so that $I(y)$ is only stationary at $y = u$ for α in this range.

13.3 Problems

13.3.1 Use the foregoing procedure to find an approximation to the solution of

$$\alpha u''(x) + (1 + x^2)u(x) = x^2 \quad \text{in} \quad 0 < x < 1$$

with

$$u'(0) = 0, \quad u(1) = 1.$$

Choose the approximating functions with care, taking advantage of what you learned from the problem leading to Figure 13.1. Compare the results and the required labor with those of other methods we have studied. How does the value of α affect your decisions and your comparisons?

13.3.2 Let

$$[p(x)u'(x)]' + q(x)u(x) = f(x) \quad \text{in} \quad a < x < b$$

with $f(a) = f'(b) = 0$.

In using the foregoing method in an effort to get a *very* accurate answer, what elementary approximation functions $\phi_k(x)$ might be used effectively in the approximation

$$u = \sum_{n=0}^{N} a_n \phi_n(x)?$$

How might you choose N?

13.3.3 Does

$$\lambda = \frac{\int_a^b \{p(x)[u'(x)]^2 - q(x)u^2(x)\}\, dx}{\int_a^b r(x)u^2(x)\, dx}$$

have a lower bound when $p(x)$ and $r(x)$ are positive in $a \leq x \leq b$?

What is a necessary condition on $w(x)$ in order that λ be stationary when $u = w(x)$?

How many functions, $w(x)$, obey this necessary condition? What values of λ accompany these functions?

Is λ minimized by any of these functions? If so, identify the corresponding function of x.

Let $p = 1$, $r = 1$, $q = 0$, and find, explicitly, the functions which render λ stationary.

Use a polynomial approximation and the method of the foregoing section to approximate the functions $w(x)$.

13.3.4 Combine the implications of Problems 13.3.3 and 13.1.5, and identify the general but nonsingular Sturm-Liouville problem with a variational principle.

13.3.5 A simple way in which to obtain a variational principle which seems to be equivalent to *any* differential equation $L(y) = 0$ is to require $\int_a^b [L(y)]^2\, dx$ to be a minimum. Compare this approach with that of Problem 13.1.2.

Separation of Variables and Product Series Solutions of Partial Differential Equations | 14

Many of the topics we have studied in connection with eigenvalue problems and special functions are motivated by the product series solution technique for partial differential equations. We now proceed to illustrate this fact.

14.1 The Heat Equation

The partial differential equation

$$\phi_{xx} - \phi_t = 0 \tag{14.1}$$

is one which arises in a great variety of scientific contexts. A typical boundary value problem is one in which $\phi(x, t)$ obeys Equation (14.1) in a domain $0 < x < a, 0 < t < \infty$ and in which, for example

$$\phi(0, t) = \phi(a, t) = 0, \quad \phi(x, 0) = g(x), \tag{14.2}$$

where $g(x)$ is a given function.

If one sought $\phi(x, t)$ in the form

$$\phi(x, t) = \sum_{n=1}^{\infty} X_n(x) T_n(t), \tag{14.3}$$

where each term in this sum is itself a solution of Equation (14.1), he might hope that the necessity of solving the foregoing problem be replaced by the necessity of solving some ordinary differential equation problems for $X_n(x)$ and $T_n(t)$. Accordingly, we first ask whether any function of the form

$$\phi(x, t) = F(x)G(t) \tag{14.4}$$

is a solution of Equation (14.1).

Substituting Equation (14.4) into Equation (14.1), we obtain

$$F''G = G'F,$$

from which we obtain immediately

$$\frac{F''(x)}{F(x)} = \frac{G'(t)}{G(t)}. \tag{14.5}$$

If we choose some fixed value for x, say x_1, then Equation (14.5) must be satisfied at the set of points $x = x_1$, $0 < t < \infty$; that is,

$$\frac{F''(x_1)}{F(x_1)} = \frac{G'(t)}{G(t)} \quad \text{in} \quad 0 < t < \infty. \tag{14.6}$$

Since, for fixed x_1, F''/F is a fixed number, it follows that*

$$G'(t)/G(t) = \text{const} = -\lambda \tag{14.7}$$

for all values of t. Thus $G(t) = (\text{const})\, e^{-\lambda t}$. It also follows from Equation (14.5) that

$$F''(x) + \lambda F(x) = 0 \tag{14.8}$$

for all values of x, so that $F(x)$ is a linear combination of $\sin(x\sqrt{\lambda})$ and $\cos(x\sqrt{\lambda})$. Thus, any linear combination of the functions ψ and χ defined by

$$\psi(x, t) = \sin(x\sqrt{\lambda})e^{-\lambda t}, \quad \chi(x, t) = \cos(x\sqrt{\lambda})e^{-\lambda t} \tag{14.9}$$

is a solution of (14.1).

It is clear that, for each λ, the only linear combinations of ψ and χ for which the $\phi(x, t)$ of Equation (14.4) satisfies both Equation (14.1) and the homogeneous conditions of Equation (14.2) are

$$\phi(x, t) \equiv 0 \quad \text{for} \quad \lambda \neq \left(\frac{n\pi}{a}\right)^2 \tag{14.10}$$

and

$$\phi_n(x, t) = A_n e^{-\lambda_n t} \sin(x\sqrt{\lambda_n}) \quad \text{for} \quad \lambda = \lambda_n = (n\pi/a)^2.$$

That is, any product $F(x)G(t)$ which satisfies Equation (14.1) and the

---
*The minus sign in Equation (14.7) is for future convenience. At this point, λ could be positive, negative, or even complex.

homogeneous boundary conditions has one factor which is any eigenfunction, F_n, of the homogeneous problem

$$F'' + \lambda F = 0$$

with

$$F(0) = F(a) = 0.$$

The other factor, $G_n(t)$, is a solution of Equation (14.7) with $\lambda = \lambda_n$. Thus, we can hope that ϕ can be described by

$$\phi(x,t) = \sum_{n=1}^{\infty} A_n \sin(x\sqrt{\lambda_n}) e^{-\lambda_n t} \tag{14.11}$$

and, in order to ascertain whether any numbers A_n provide a function $\phi(x, t)$ that obeys all of the constraints (14.1), (14.2), we appeal again to Equation (14.2); i.e., we write

$$\phi(x, 0) = \sum_{n=1}^{\infty} A_n \sin(x\sqrt{\lambda_n}) = g(x). \tag{14.12}$$

This equation identifies the A_n as the Fourier coefficients of the eigenfunction expansion of $g(x)$. That is, we have

$$A_n = \frac{2}{a} \int_0^a g(x) \sin \frac{n\pi x}{a}\, dx,$$

and Equation (14.11) now describes $\phi(x, t)$ without ambiguity.

We illustrate an alternative, and perhaps more careful, approach in connection with another typical boundary value problem. Let

$$u_{xx} - u_t = h(x, t) \tag{14.13}$$

for $0 < x < a$, $t > 0$, and with

$$u(0, t) = u'(a, t) = u(x, 0) = 0. \tag{14.14}$$

We would like to find a solution to this problem in the form

$$u(x, t) = \sum_{n=1}^{\infty} f_n(t) \cdot w_n(x). \tag{14.15}$$

If the $w_n(x)$ are eigenfunctions of *some* Sturm-Liouville problem associated with the interval $(0, a)$, then (thinking of t as a parameter) the validity of Equation (14.15) is guaranteed by the fundamental expansion theorem of Sturm-Liouville theory. There is, of course, a very natural Sturm-Liouville problem to use in generating the eigenfunctions $w_n(x)$;

it is that which arises in the method of separation of variables (cf. Equation (14.8)) as applied to the homogeneous counterpart of Equation (14.13), with the associated boundary conditions $w_n(0) = w'_n(a) = 0$ as suggested by Equation (14.14). Thus, we choose the $w_n(x)$ as the eigenfunctions of the problem

$$w''_n + \sigma_n w_n = 0, \quad w_n(0) = w'_n(a) = 0 \tag{14.16}$$

and we now know that the expansion (14.15) must be possible.

To determine the functions $f_n(t)$, we substitute Equation (14.15) into Equation (14.13) to obtain*

$$\sum_{n=1}^{\infty} [f_n(t)w''_n(x) - f'_n(t)w_n(x)] = h(x, t). \tag{14.17}$$

If we use Equation (14.16) and also write

$$h(x, t) = \sum_{n=1}^{\infty} h_n(t) \, w_n(x),$$

Equation (14.17) becomes

$$\sum_{n=1}^{\infty} [-\sigma_n f_n(t) - f'_n(t)] w_n(x) = \sum_{n=1}^{\infty} h_n(t) w_n(x),$$

so that

$$f'_n(t) + \sigma_n f_n(t) = -h_n(t). \tag{14.18}$$

Since $u(x, 0) = 0$, Equation (14.15) requires that $f_n(0) = 0$; we can now solve Equation (14.18) subject to this initial condition so as to obtain the $f_n(t)$ uniquely.

14.2 Problems

14.2.1 Complete the solution of Equations (14.1) and (14.2) when

(a)
$$g(x) = x(a - x);$$

(b)
$$g(x) = \begin{cases} x, & 0 < x < a/2, \\ a - x, & a/2 < x < a; \end{cases}$$

(c)
$$g(x) = \delta\left(x - \frac{a}{2}\right).$$

Compare the results of these Problems.

---
*Note that term-by-term differentiation has been assumed to be permissible. When dealing with infinite series, this may not always be the case; cf. Problem (14.4.5).

14.2.2 Complete the solution of Equations (14.13) and (14.14) when

(a)
$$h(x, t) = \begin{cases} 1, & 0 < x < a, \ 0 < t < T, \\ 0, & t > T; \end{cases}$$

(b)
$$h(x, t) = \delta(x - \beta t).$$

Compare these results.

14.2.3 Examine the consequences of choosing the $w_n(x)$ in Equation (14.15) to be eigenfunctions of some Sturm-Liouville problem other than that of Equation (14.16).

The foregoing problems have characterizing features on which we shall build further. They are:

(1) a linear partial differential equation whose homogeneous form admits solutions of the form $\phi(x, y) = F(x)G(y)$;

(2) a domain $x_1 < x < x_2$, $y_1 < y < y_2$;

(3) homogeneous boundary conditions on x_1 and x_2 (or, alternatively, on y_1 and y_2).

14.3 Nonhomogeneous Boundary Conditions

There is an elementary device whereby problems with nonhomogeneous boundary conditions can be recast in a form with homogeneous boundary conditions. To illustrate this we let

$$u_{xx} - u_t = 0 \quad \text{in} \quad 0 < x < 1, \ 0 < t < \infty \quad (14.19)$$

with

$$u(0, t) = 0 \quad u(1, t) = e^{-\alpha t}, \quad u(x, 0) = x^2,$$

and we define

$$u = x^2 e^{-\alpha t} + w(x, t). \quad (14.20)$$

Substituting Equation (14.20) into Equation (14.19), we obtain

$$w_{xx} - w_t = -(2 + \alpha x^2)e^{-\alpha t} \quad (14.21)$$

with

$$w(0, t) = w(1, t) = w(x, 0) = 0. \quad (14.22)$$

The determination of w by a series expansion is now straightforward.

14.4 Problems

14.4.1 Complete the solution of Equations (14.21) and (14.22) according to the foregoing method.

14.4.2 Solve the Equation (14.19) by defining

$$u(x, t) = h(x)e^{-\alpha t} + y(x, t),$$

where $h(x)$ is to be chosen so that

$$y_{xx} - y_t = 0 \quad \text{in} \quad 0 < x < 1, \quad 0 < t < \infty$$

with

$$y(0, t) = y(1, t) = 0.$$

You will have to identify the boundary condition on y at $t = 0$ as part of the procedure.

14.4.3 Use the results of Problems 14.4.1 and 14.4.2 to characterize the dependence of the solution of (14.19) on α.

14.4.4 By constructing simple examples, verify that unless the three characterizing features listed just before Section 14.3 are met, the product series solution technique runs into trouble.

14.4.5 In connection with the problem of Equations (14.13) and (14.14), we used functions $w_n(x)$ satisfying the same homogeneous boundary conditions at $x = 0$ and $x = a$ as those satisfied by $u(x, t)$. If $u(x, t)$ does not satisfy homogeneous boundary conditions, then, even though the expansion (14.15) using our previous $w_n(x)$ is still feasible in principle, we can anticipate that the convergence of this series is not as strong as before; in fact, the series obtained by differentiation (as in Equation (14.17)) may not converge. An alternative procedure would be to determine the coefficients $f_n(t)$ by use of the formula *

$$f_n(t) = \int_0^a u(x, t) w_n(x)\, dx.$$

Multiplication of Equation (14.13) by $w_n(x)$ and integration by parts then leads to a differential equation for $f_n(t)$ in which the boundary conditions for $u_n(x, t)$ appear. Carry through this *finite transform* process for the problem of Equation (14.19), and compare your results with those for Problem 14.4.1.

---

* We here assume the $w_n(x)$ to be normalized.

14.4.6 Obtain the Laplace transform (in time) of $\phi(x,t)$ satisfying Equations (14.1) and (14.2), and use Equation (14.11) to deduce a $f(t) \rightleftharpoons F(s)$ formula.

14.5 Laplace's Equation

Although boundary-value problems in which the heat equation appears are extremely useful for the introduction of the method of separation of variables, it is almost always true that integral transform methods provide a better way* of treating such problems. This is not true of the boundary-value problem involving *Laplace's equation*

$$u_{xx} + u_{yy} = 0 \quad \text{in} \quad 0 < x < a, \; 0 < y < b \qquad (14.23)$$

with, for example,

$$u(0, y) = u(a, y) = u(x, 0) = 0, \quad u(x, b) = g(x). \qquad (14.24)$$

Since the boundary conditions at $x = 0$ and at $x = a$ are both homogeneous, we should now expect that the solution of Equations (14.23) and (14.24) might admit a description

$$u(x, y) = \sum_{n=1}^{\infty} Y_n(y) X_n(x), \qquad (14.25)$$

where the $X_n(x)$ are eigenfunctions of an appropriate eigenfunction problem. Accordingly, we seek product series solutions of Equation (14.23) by substituting $Y(y)X(x)$ for u in Equation (14.23). We obtain

$$\frac{Y''(y)}{Y(y)} = -\frac{X''(x)}{X(x)}$$

and, using the arguments of the foregoing section, we conclude that each side of this equation is a constant, λ, so that

$$X''(x) + \lambda X = 0 \qquad (14.26)$$

and

$$Y''(y) - \lambda Y = 0. \qquad (14.27)$$

Using the boundary conditions at $x = 0$ and $x = a$ as given in Equation

---
* Provided the user of the method is conversant with complex variable theory and techniques.

(14.24), we conclude that $X(x)$ vanishes identically unless $\lambda = \lambda_n = (n\pi/a)^2$; the eigenfunction corresponding to λ_n is

$$X_n(x) = \sin(x\sqrt{\lambda_n}) = \sin\frac{n\pi x}{a}. \tag{14.28}$$

We now follow the same kind of approach as that outlined in connection with Equation (14.15). We write

$$u(x, y) = \sum_{n=1}^{\infty} a_n(y) \sin\frac{n\pi x}{a} \tag{14.29}$$

where the existence of such an expansion is guaranteed by the fact that the functions $\sin(n\pi x/a)$ form a complete set of eigenfunctions. Substitution into Equation (14.23) leads to

$$\sum_{n=1}^{\infty} \left[-\frac{n^2\pi^2}{a^2} a_n(y) + a_n''(y) \right] \sin\frac{n\pi x}{a} = 0, \tag{14.30}$$

so that

$$a_n(y) = A_n \cosh\frac{n\pi y}{a} + B_n \sinh\frac{n\pi y}{a}. \tag{14.31}$$

The boundary conditions (14.24) imply that

$$A_n = 0, \quad \sum_{n=1}^{\infty} \left(B_n \sinh\frac{n\pi b}{a} \right) \sin\frac{n\pi x}{a} = g(x) = \sum_{n=1}^{\infty} g_n \sin\frac{n\pi x}{a}, \tag{14.32}$$

where g_n is the expansion coefficient of $g(x)$. Thus, we have

$$B_n = \frac{g_n}{\sinh(n\pi b/a)}$$

and

$$u(x, y) = \sum_{n=1}^{\infty} g_n \frac{\sinh(n\pi y/a)}{\sinh(n\pi b/a)} \sin\frac{n\pi x}{a}. \tag{14.33}$$

Had we started with a *Poisson equation*,

$$u_{xx} + u_{yy} = h(x, y), \tag{14.34}$$

instead of Equation (14.23), then Equation (14.30) would have been replaced by

$$\sum_{n=1}^{\infty} \left[-\frac{n^2\pi^2}{a^2} a_n(y) + a_n''(y) \right] \sin\frac{n\pi x}{a} = \sum_{n=1}^{\infty} h_n(y) \sin\frac{n\pi x}{a},$$

where $h_n(y)$ is the expansion coefficient of $h(x, y)$; we would thus have had to solve the equation

$$a_n''(y) - \frac{n^2\pi^2}{a^2} a_n(y) = h_n(y).$$

The arbitrary constants appearing in the solution would be determined by use of the boundary conditions (14.24), as before.

Finally, had the boundary conditions satisfied by $u(x, y)$ at $x = 0$ and at $x = a$ not been homogeneous, then to avoid the probable nonconvergence of the differentiated series appearing in Equation (14.30), we would have had to make a preliminary modification in the problem by defining a new function ϕ which differs from u by an additive function of x and y, so chosen that ϕ does satisfy the same homogeneous conditions at $x = 0$ and $x = a$ as are satisfied by the $w_n(x)$. (Alternatively, we could have used a method similar to that of Problem 14.4.5.)

14.6 Problems

14.6.1 In the problem of Equations (14.23) and (14.24), let

$$g(x) = x,$$

evaluate the coefficients g_n, and calculate enough of the terms in Equation (14.33) so that you can characterize, thoroughly, the behavior of $u(x, y)$ when $b/a \gg 1$, when $b/a \ll 1$, and when $b/a \cong 1$.

14.6.2 Use a product series expansion to solve

$$q_{xx} + q_{yy} = xe^y \quad \text{in } 0 < x < a, \ 0 < y < b,$$

where

$$q(0, y) = q(x, 0) = 0, \quad q(a, y) = y, \quad q(x, b) = \frac{xb}{a} + \exp(-1/[x(a-x)]).$$

14.6.3 Let

$$u_{xx} + u_{yy} = 0 \quad \text{in } 0 < x < 2, \ 0 < y < 2,$$

and let

$$u(0, y) = u(2, y) = u(x, 0) = 0$$

and

$$u(x, 2) = \begin{cases} 1, & x < 1, \\ 0, & x > 1. \end{cases}$$

Find $u(1, 1)$ without using product series or any other technique requiring extensive analysis or calculation.

14.6.4 Find a product series solution of

$$u_{xx} + u_{yy} = \delta(x - \xi)\,\delta(y - \eta)$$

in the square region $-1 < x < 1$, $-1 < y < 1$, where $u = 0$ on all boundaries, and where the point (ξ, η) is inside the square.

14.7 Another Geometry

Suppose now that

$$Z_{xx} + Z_{yy} = 0 \quad \text{in} \quad x^2 + y^2 < 1 \tag{14.35}$$

and that

$$Z(x, \pm\sqrt{1-x^2}) = x^2. \tag{14.36}$$

We have already learned that Equation (14.35) admits solutions of the form $X(x)Y(y)$, but the boundary conditions are not given on segments $x = $ constant and $y = $ constant. Therefore, although one can hope and expect (correctly) that Z can be described by

$$Z = \sum X_n(x) Y_n(y),$$

any hope that a direct process of the foregoing variety will identify $X_n(x)$ and $Y_n(y)$ is not justified!

However, we recall that

$$Z_{xx} + Z_{yy} \equiv \text{div grad}\,[Z(x, y)],$$

and that, in the polar coordinate system in which $x = r\cos\theta$, $y = r\sin\theta$, we have

$$\text{div grad}\,w(r, \theta) = \frac{1}{r}(rw_r)_r + \frac{1}{r^2} w_{\theta\theta}.$$

Thus, if we decide to describe the values of $Z(x, y)$ in terms of polar coordinates, we can define

$$w(r, \theta) \equiv Z(x, y)$$

and, since div grad $Z = 0$, we have

$$\frac{1}{r}(rw_r)_r + \frac{1}{r^2} w_{\theta\theta} = 0 \quad \text{in} \quad r < 1 \tag{14.37}$$

and

$$w = \cos^2\theta \quad \text{on} \quad r = 1. \tag{14.38}$$

We must note immediately that corresponding to each point x, y in $r < 1$ there is one value of r (that is, $r = \sqrt{x^2 + y^2}$), but there are many values of θ. In particular, we can decide to confine our attention to values of θ such that $0 \leq \theta \leq 2\pi$, in which case, if w is to be continuous and differentiable throughout the region $r < 1$, we must require that

$$w(r, 0) = w(r, 2\pi), \qquad w_\theta(r, 0) = w_\theta(r, 2\pi). \tag{14.39}$$

The reader should verify that Equations (14.37) and (14.39) imply that

$$\frac{\partial^n w}{\partial \theta^n}(r, 0) = \frac{\partial^n w}{\partial \theta^n}(r, 2\pi)$$

with $n > 1$.

Alternatively, we can decide to describe $w(r, \theta)$ for all real θ. If we do so, we must accept the implication that, since Z (and hence w) has only one value at a given x, y, then $w(r, \theta)$ must be periodic in θ with period 2π; that is, we have

$$w(r, \theta) = w(r, \theta + 2\pi) \tag{14.40}$$

for all θ.

The two alternatives *must* lead to a single result, since they are merely alternative views of the same problem. We adopt the latter alternative and face the problem which requires that we solve Equation (14.37) subject to Equation (14.38) and (14.40). The boundary is now a curve along which one of the independent variables is constant and we can hope to use separation of variables. Thus, we ask whether any function

$$W = R(r)\Theta(\theta) \tag{14.41}$$

will satisfy Equation (14.38).

We substitute Equation (14.41) into Equation (14.37) and obtain

$$\frac{r(rR')'}{R} = -\frac{\Theta''}{\Theta}$$

and conclude, as in all of our separation of variable problems, that

$$r(rR')' - \lambda R = 0, \tag{14.42}$$

$$\Theta'' + \lambda \Theta = 0, \tag{14.43}$$

where λ is a constant.

The function Θ and the appropriate values of λ are determined by Equations (14.43) and (14.40)—a Sturm-Liouville problem with periodic end conditions. It is evident that, when $\lambda = n^2$, both of the linearly independent solutions of Equation (14.43) are periodic with period 2π

and that, when $\lambda \neq n^2$, Equation (14.43) has no solutions which have period 2π. Therefore, we will try to construct a solution involving products of the functions $\cos n\theta$ and $\sin n\theta$ (or $e^{in\theta}$ and $e^{-in\theta}$) and the functions $R_n(r)$ which are solutions of Equation (14.42) with $\lambda = n^2$. These solutions are (as the reader has demonstrated if he has done the problems of earlier chapters),

$$R_n(r) = A_n r^n + B_n r^{-n} \quad (n > 0), \qquad R_0(r) = A_0 + B_0 \ln r. \quad (14.44)$$

Unless $B_n = 0$ for each n, the function $R_n(r)\Theta_n(\theta)$ will not be defined at $r = 0$. Since $r = 0$ is a point at which the solution must not only be defined but must be twice differentiable (why?), B_n is zero and the only surviving products lead to the following description of $w(r, \theta)$:

$$w(r, \theta) = \sum_{n=0}^{\infty} A_{1n} r^n \cos n\theta + \sum_{n=1}^{\infty} A_{2n} r^n \sin n\theta, \quad (14.45)$$

where the sums have the same interpretation they have had for more conventional eigenfunction expansions, and where the A_{1n} and A_{2n} must be determined by the boundary condition

$$\sum_{n=0}^{\infty} A_{1n} \cos n\theta + \sum_{n=1}^{\infty} A_{2n} \sin n\theta = \cos^2 \theta \equiv \frac{1 + \cos 2\theta}{2}. \quad (14.46)$$

Thus, we have

$$A_{10} = \tfrac{1}{2}, \qquad A_{12} = \tfrac{1}{2},$$

and each other coefficient is zero.

It follows directly that

$$w(r, \theta) = \frac{1 + r^2 \cos 2\theta}{2} = \frac{1 - r^2}{2} + x^2.$$

It is clear that this solution could have been found without the product series expansion technique, but that was not our purpose.

14.8 Problems

14.8.1 Use a product series expansion method to find $w(r, \theta)$ satisfying

$$\frac{1}{r}(rw_r)_r + \frac{1}{r^2} w_{\theta\theta} = r^3 e^{r\theta}$$

in the sectorial region $1 < r < 2$, $0 < \theta < \pi/3$, where $w(1, \theta) = 0$, $w(2, \theta) = 1$, $w(r, 0) = r - 1$, $w(r, \pi/3) = (r - 1)^3$. Do this by use of two different expansions, one based on eigenfunctions in the r-variable and the other on eigenfunctions in the θ-variable.

14.9 The Helmholtz Equation

Sections (14.1) and (14.5) have been chosen to illustrate the manner in which the *separation of variables* plus *product series expansion* technique is used. Clearly, no unfamiliar special functions have been displayed; neither have exhaustive studies of Laplace's equation and the heat equation been attempted. In this section we encounter still another important equation of mathematical physics which will lead us to the special functions which give Chapter 11 its name.

In a form which is not tied to any particular coordinate system, Helmholtz's equation is

$$\text{div grad } \psi + \beta^2 \psi = 0, \qquad (14.47)$$

where ψ is a function of three coordinate variables. It is convenient to describe ψ in terms of coordinates which are chosen to fit the geometry of the problem. For example, let ψ be required to satisfy Equation (14.47) in the spherical domain $x^2 + y^2 + z^2 < R^2$ with given values of ψ on each point of $x^2 + y^2 + z^2 = R^2$. The appropriate coordinate system is the spherical one for which

$$x = r \sin \theta \cos \phi, \qquad y = r \sin \theta \sin \phi, \qquad z = r \cos \theta, \qquad (14.48)$$

and the domain lies in

$$0 \leq r < R, \qquad 0 < \theta < \pi, \qquad 0 < \phi < 2\pi.$$

When ψ is described in terms of the variables (14.48), Equation (14.47) takes the form

$$(r^2 \psi_r)_r + \frac{1}{\sin \theta} (\sin \theta \, \psi_\theta)_\theta + \frac{1}{\sin^2 \theta} \psi_{\phi\phi} + \beta^2 r^2 \psi = 0. \qquad (14.49)$$

Our present purposes are best served by limiting our attention to problems in which ψ depends only on r and θ. In fact, we admit for consideration here only boundary conditions of the form

$$\psi(R, \theta, \phi) = g(\theta). \qquad (14.50)$$

With this restriction, Equation (14.49) simplifies to

$$(r^2 \psi_r)_r + \frac{1}{\sin \theta} (\sin \theta \, \psi_\theta)_\theta + \beta^2 r^2 \psi \equiv 0, \qquad (14.51)$$

and the substitutions

$$x = \cos \theta, \qquad \psi(r, \theta) = \Phi(r, x)$$

provide, as an alternative to Equation (14.51), the equation

§14.9] The Helmholtz Equation 143

$$(r^2\Phi_r)_r + [(1 - x^2)\Phi_x]_x + \beta^2 r^2 \Phi = 0. \tag{14.52}$$

The reader can verify that functions, Φ, of the form

$$\Phi = F(r)G(x)$$

are solutions of Equation (14.52), provided that

$$(r^2 F')' + \beta^2 r^2 F - kF = 0 \tag{14.53}$$

and

$$[(1 - x^2)G']' + kG = 0. \tag{14.54}$$

We know, from the problems of Chapter 11, that Equation (14.53) leads to Bessel functions, and that Equation (14.54) is Legendre's equation.

We consider now the specific boundary-value problem in which

$$\Delta\psi + \beta\psi = 0 \quad \text{in } r < R, \ 0 \le \theta \le \pi, \ 0 \le \phi \le 2\pi$$

with

$$\psi(R, \theta, \phi) = \begin{cases} 0, & 0 \le \theta \le \frac{\pi}{3}, \\ 1, & \frac{\pi}{3} \le \theta \le \frac{2\pi}{3}, \\ 0, & \frac{2\pi}{3} \le \theta < \pi. \end{cases}$$

We seek $\psi(r, \theta, \phi)$ in the form

$$\psi = \sum_{n=0}^{\infty} F_n(r)G_n(x), \tag{14.55}$$

where $x = \cos\theta$. We know that, when $k = n(n + 1)$, the solution $G_n(x)$ of Equation (14.54) is

$$B_n(x) = P_n(x),$$

where the functions, P_n, are the Legendre polynomials. For other values of k, $G(x)$ is unbounded at $x = \pm 1$ and $F(r)G(x)$ is not defined in part of the region in which Equation (14.51) must be satisfied. Thus, only if $\psi(R, \theta, \phi)$ can be written as

$$\psi(R, \theta, \phi) = \sum_{n=0}^{\infty} F_n(r)P_n(x) \tag{14.56}$$

can Equation (14.55) provide a valid description of ψ. Fortunately, the functions $P_n(x)$ form a complete set over the interval $(-1, 1)$ (a fact we use without proof). Using the properties of the functions $P_n(x)$ given in Section (11.5), it follows that

$$\frac{2}{2n+1} F_n(R) = \int_{-1}^{1} \psi(R, \theta, \phi) P_n(x)\, dx \tag{14.57}$$

with $\cos \theta = x$.

We must now find those $F_n(r)$ for which Equation (14.53) holds in $r < R$, with an appropriate boundary condition. The reader can verify that

$$\psi = \sum_{n=0}^{\infty} F_n(R) \left(\frac{r}{R}\right)^{-\frac{1}{2}} \frac{J_{n+\frac{1}{2}}(\beta r)}{J_{n+\frac{1}{2}}(\beta R)} P_n(x), \tag{14.58}$$

where

$$\frac{2}{2n+1} F_n(R) = \int_{-\frac{1}{2}}^{\frac{1}{2}} P_n(x)\, dx. \tag{14.59}$$

Is this solution acceptable in view of the factor $(r/R)^{-\frac{1}{2}}$ in each term of the series?

Nonlinear Differential Equations | 15

To illustrate how a first-order nonlinear differential equation might arise, consider the rate of population growth of a short-lived parasite. Let the parasite population at t be $P(t)$, and let the total food supply be $F(t)$. We postulate that the birth rate is proportional to the number of pairs, and so to $P^2(t)$, and also to the food supply per capita, namely F/P; the death rate we take to be proportional to P. Thus, we have

$$\frac{dP}{dt} = aP^2 \frac{F}{P} - kP, \tag{15.1}$$

where a and k are positive constants. We postulate also that the food supply alters according to the equation

$$\nu \frac{dF}{dt} = c - F - \beta P; \tag{15.2}$$

that is, F would differ from c by an amount $Ae^{-t/\nu}$ and, hence, would tend to stabilize at the level c, except that it is devoured at a rate proportional to P. Here ν, c, β, and A are positive constants.

Dividing Equation (15.1) by Equation (15.2), we obtain

$$\frac{dP}{dF} = \frac{\nu P(aF - k)}{c - F - \beta P}, \tag{15.3}$$

which is a first-order nonlinear equation. An equation like (15.3) is not easy to solve analytically; recourse to graphical or numerical methods is usually necessary.

Consider a plot of P against F (Figure 15.1). The slope dP/dF of a solution curve must be infinite whenever that curve intersects the straight line $c - F - \beta P = 0$, and it must be zero whenever it intersects the vertical line $F = k/a$. For a point (F, P) above the sloping line, dP/dF will be > 0 or < 0 according as $F < k/a$ or $F > k/a$, respectively. Thus solution curves (which are, of course, constrained to the first quadrant) must be of the kind sketched in the figure. For equilibrium, $dP/dt = dF/dt = 0$, and Equations (15.1) and (15.2) show that this can occur only at the points $(c, 0)$ and $(k/a, (c - k/a)/\beta)$, marked A and B, respectively.

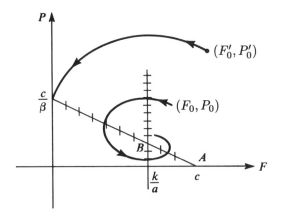

Arrows indicate direction of increasing t.

Figure 15.1 Parasite problem.

Suppose that the values of F and P at $t = 0$ are F_0 and P_0, respectively. Then, in general [however, cf. Problem (15.5.2)], we can anticipate that, as time passes, the solution curve will spiral in around point B, finally approaching it as a limit. If, however, P_0 is sufficiently large, then the curve will intersect the P-axis, that is, F will vanish. If F itself represents a living organism, perhaps one which it is desirable to exterminate by supplying a sufficiently large amount of P_0, then the minimum value of P_0 required for such extermination must correspond to the point (F_0', P_0') lying on that solution curve passing through the point $(0, c/\beta)$.

15.1 Problems

15.1.1 Choose reasonable values for the constants in the above example, and sketch the solution curves with some care. For $F_0 = c$, estimate the minimum value of P_0 required for extinction of F.

[In plotting solution curves, it is useful to observe that the slope dP/dF has the constant value α on that curve in the FP plane (called an *isocline*) satisfying

$$\frac{dP}{dF} = \frac{\nu P(aF - k)}{c - F - \beta P} = \alpha.]$$

In general, which parameter groups determine the efficiency of a particular "specific" in eliminating pests?

15.1.2 Sometimes it is possible to transform a nonlinear differential equation into a linear one by the use of an appropriate device. Solve or simplify each of the following examples, which are of this character. (Several are of historical interest. Unfortunately, they are rarely encountered in practice!)

(a) *Homogeneous* equation: $y' = f(y/x)$, where f is some prescribed function. (Set $y/x = u$, and separate variables.)

(b) If a is a constant, determine the family of solution curves for $y' = ax/y$. Find also their orthogonal trajectories.

(c) $y' = g[(ax + by + c)/(dx + ey + f)]$, where a, b, c, d, e, f are constants and g is a prescribed function. Discuss also the special case $ae = bd$.

(d) A point moves in the xy-plane so that its distance from the origin is equal to the distance from the origin to the intersection point of the y-axis with the tangent drawn to the curve generated by the moving point. Find the equation of that curve.

(e) Bernoulli equation: $y' + f(x)y + y^c g(x) = 0$ where $f(x), g(x)$ are prescribed functions and c is a constant. (Set $u = y^\alpha$ for an appropriate α.)

(f) Riccati equation: $y' + f(x)y + y^2 g(x) = h(x)$. (Set $u'/u = gy$.) Compare what you have just done with the material of Chapter 10.

(g) $y = f(y')$, where f is a prescribed function. (Set $y' = u$, and treat u as a parameter, expressing each of x, y in terms of u.)

(h) Use the technique of (g) for the Clairaut equation $y = xy' + f(y')$, and for the d'Alembert equation $y = xg(y') + f(y')$.

15.1.3 Plot the solution curves of

$$y' = x^2 - y^2 \tag{15.4}$$

in the xy-plane, using either the method of isoclines (cf. Problem 15.1.1) or any other convenient and rapid graphical method. Demonstrate graphically that if $y(0)$ is greater than some lower bound α (estimate α), the solution for $x > 0$ will eventually turn upwards and become asymptotic to the line $y = x$, whereas if $y(0) < \alpha$, the solution curve will proceed downward with increasing steepness.

15.1.4 Solve Equation (15.4) exactly, using the transformation $y = u'/u$ followed by $u = \sqrt{x} f(\tfrac{1}{2} x^2)$ to turn it into a form of Bessel's equation. In particular, show that

$$y = x \frac{A I_{-\frac{1}{4}}(x^2/2) + I_{\frac{1}{4}}(x^2/2)}{A I_{\frac{1}{4}}(x^2/2) + I_{-\frac{1}{4}}(x^2/2)}, \tag{15.5}$$

where $A = \tfrac{1}{2} y(0) \Gamma(\tfrac{1}{4}) / \Gamma(\tfrac{3}{4})$. Using the fact that the modified Bessel function $I_\nu(x) \sim e^x/\sqrt{2\pi x}$ for large $x > 0$, determine the asymptotic behavior of y. Is your result compatible with the "α situation" described in Problem 15.1.3?

15.2 Second-Order Equations; The Phase Plane

Nonlinear equations are usually difficult to handle, and recourse to numerical or approximate methods is frequently necessary. It may happen, however, that the equation in question, in some sense, is "close" to a linear equation; the perturbation techniques of Chapters 9 and 18 may then be applicable. We consider here another situation in which certain second-order nonlinear equations can be treated fairly simply, that in which the equation has the form

$$f(y, y', y'') = 0, \tag{15.6}$$

where f is some prescribed function. Note that the independent variable, t, does not occur in Equation (15.6). The idea now is to consider solution curves for Equation (15.6) in the yy'-plane (called the *phase plane*) rather than in the usual *ty*-plane. While a solution curve in the *yy'*-plane does not provide *y* as a function of *t* directly, it is an easy matter to determine *t* for any point on the curve by integrating the relation $dt = dy/y'$ along the curve.

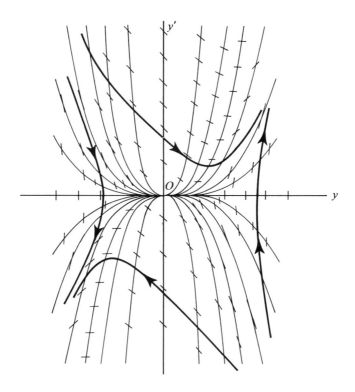

Figure 15.2 Isoclines for $y'' = -y' + y^3$.

To show that Equation (15.6) provides a description of solution curves in the phase plane, we need merely note that

$$y'' = \frac{dy'}{dy} \cdot y', \tag{15.7}$$

so that Equation (15.6) becomes

$$f\left(y, y', y'\frac{dy'}{dy}\right) = 0. \tag{15.8}$$

Equation (15.8) gives a formula for the slope dy'/dy of a solution curve passing through the point (y, y').

As an example, consider the equation

$$y'' + y' = y^3 \tag{15.9}$$

which becomes

$$y'\frac{dy'}{dy} + y' = y^3. \tag{15.10}$$

In plotting the curves satisfying Equation (15.10) in the phase plane, it is convenient to sketch first the isoclines (i.e., the lines on which dy'/dy has a constant value α) and to draw numerous short lines of slope α on each such isocline. This has been done in Figure 15.2, where the isoclines have been drawn. Some typical solution curves (called *trajectories*) are also drawn. Arrows indicate the direction of increasing t.

Alternatively, a simple geometrical construction generating the slope dy'/dy at any point (y, y') can sometimes be devised, with the exercise of some ingenuity.

Sometimes we can obtain the equations of the phase plane trajectories directly. This is the case, for example, for the nonlinear oscillator problem described by

$$\frac{d^2x}{d\tau^2} + \omega^2 x = \epsilon x^2, \tag{15.11}$$

where ω and ϵ are given positive constants. Here x is position and τ is time; the presence of the term ϵx^2 results in a distortion of the otherwise sinusoidal motion. Rather than plot a new set of phase plane trajectories for each choice of the pair (ω, ϵ), we first set $\tau = t/\omega$ and $x = \omega^2 y/\epsilon$ to obtain

$$\frac{d^2y}{dt^2} + y = y^2.$$

Termwise multiplication by dy/dt and integration yield

$$\frac{1}{2}\left(\frac{dy}{dt}\right)^2 + \frac{1}{2}y^2 = \frac{1}{3}y^3 + k, \tag{15.12}$$

where k is a constant of integration. The corresponding phase-plane plot is shown in Figure 15.3.

The reader should convince himself that the points $(0, 0)$ and $(1, 0)$ are equilibrium points in the sense that the representative point will not depart from either of these points if placed there initially, that the time required to approach the point $(1, 0)$ along a trajectory through it is infinite, and that, in terms of the original variables, the motion will be oscillatory if $|\dot{x}(0)| < \omega^3/(\sqrt{3}\epsilon)$.

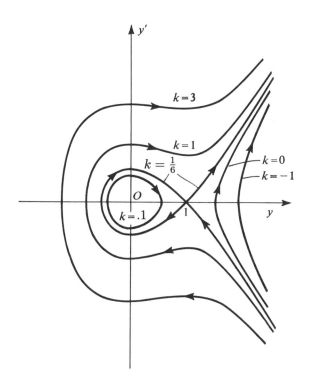

Figure 15.3 Trajectories for Equation (15.12).

15.3 Problems

15.3.1 Sketch the phase plane trajectories for the equations

(a) $\ddot{x} + \omega^2 x = -\epsilon x$,

(b) $\ddot{x} + \omega^2 x = \epsilon x^3$

where ω and ϵ are constants. Attach arrows to the trajectories.

15.3.2

(a) The equation governing the motion of a simple pendulum without damping is

$$\ddot{\theta} + \frac{g}{l}\sin\theta = 0, \qquad (15.13)$$

where g is the acceleration of gravity and l is the length of the pendulum arm; θ is the angle between the pendulum arm and the (downward) vertical. Sketch the phase plane trajectories: include the case in which the bob can go "over the top."

(b) Add a damping term $(-C\dot{\theta}|\dot{\theta}|)$ to the right hand side of Equation (15.13) and plot the new trajectories; here C is a positive constant.

15.3.3 An often-discussed equation describing a certain kind of oscillatory systems is that of van der Pol:

$$\ddot{x} + \epsilon(x^2 - 1)\dot{x} + x = 0, \qquad (15.14)$$

where ϵ is a constant. Plot the phase plane trajectories for each of three cases: (a) $\epsilon = 2$, (b) large $\epsilon > 0$, and (c) small $\epsilon > 0$. Are there any situations in which it appears that the oscillations will tend toward a closed curve (*limit cycle*) in phase space? Can you think of a simple system in economics, physics, or some other discipline which could be reasonably described by Equation (15.14)?

15.4 Singular Points

In the preceding section, we studied equations of form $f(y, y', y'') = 0$ or, equivalently,

$$\frac{d^2y}{dt^2} = F\left(y, \frac{dy}{dt}\right). \qquad (15.15)$$

If we define $z = dy/dt$, Equation (15.15) may be replaced by the equations

$$\frac{dy}{dt} = z, \qquad \frac{dz}{dt} = F(y, z).$$

More generally, we will consider the system

$$\frac{dz}{dt} = F(y, z), \qquad (15.16)$$

$$\frac{dy}{dt} = G(y, z), \qquad (15.17)$$

where F and G are prescribed functions. (Note, incidentally, that Equations (15.1) and (15.2) are of this form.) Since t does not occur explicitly in either F or G, a (y, z) phase plane approach is still appropriate. Our present concern is with the case in which each of F and G vanishes at a certain point (y_0, z_0), termed a *singular point* of the equation set. In such a case, dz/dy is not determined, so that the slope of a solution curve through the phase plane point (y_0, t_0) cannot be calculated. (Observe that the case $F(y_0, z_0) \neq 0$, $G(y_0, z_0) = 0$ is not troublesome, since the roles of y and z may simply be interchanged.)

Let us now analyze the behavior of the system (15.16), (15.17) near an isolated singular point (y_0, z_0); without loss of generality, we can translate the coordinate system so that this point is the origin. Thus we start with the assumption that $F(0, 0) = G(0, 0) = 0$. We postulate also that each of F and G is well-behaved, and so, for example, can be differentiated with respect to each of y and z as many times as is necessary.

An immediate observation is that the origin is an equilibrium (stationary) point, since not only $dz/dt = dy/dt = 0$ there, but also $d^2z/dt^2 = d^2y/dt^2 = 0$, $d^3z/dt^3 = d^3y/dt^3 = 0$, etc., as a result of repeated differentiation of Equations (15.16) and (15.17). Thus if the representative point in the phase plane is placed at the origin to start with, it will always remain there. Consider next the immediate neighborhood of the origin. Here we use Taylor's theorem to approximate F and G with

$$F \cong \alpha y + \beta z, \qquad (15.18)$$

$$G \cong \gamma y + \delta z, \qquad (15.19)$$

where α, β, γ, and δ are constants; we restrict ourselves to the case in which not all of α, β, γ, δ vanish. It is reasonable to expect that the motion of the system (15.16), (15.17) near the origin is adequately described * by

$$\frac{dz}{dt} = \alpha y + \beta z, \qquad (15.20)$$

$$\frac{dy}{dt} = \gamma y + \delta z. \qquad (15.21)$$

We can easily solve this system of equations, subject to any given initial conditions prescribing $z(0)$ and $y(0)$. One method is to use Laplace

---

* Indeed it is, except perhaps for consideration of stability. A thorough discussion will be found in L. A. Pars, *A Treatise on Analytical Dynamics*, Heinemann, 1965, Chapter 19.

transforms; denoting by $Z(s)$ and $Y(s)$ the transforms of $z(t)$ and $y(t)$ and taking the transform of Equations (15.20) and (15.21), we obtain

$$Z(s) = \frac{\alpha y(0) + (s - \gamma)z(0)}{s^2 - s(\beta + \gamma) - (\alpha\delta - \beta\gamma)}, \qquad (15.22)$$

$$Y(s) = \frac{(s - \beta)y(0) + \delta z(0)}{s^2 - s(\beta + \gamma) - (\alpha\delta - \beta\gamma)}. \qquad (15.23)$$

These transform expressions may be inverted by the usual partial fraction expansion process (cf. Chapter 12); it is clear that the behavior of $z(t)$ and $y(t)$ will depend on the roots of the equation

$$s^2 - s(\beta + \gamma) - (\alpha\delta - \beta\gamma) = 0. \qquad (15.24)$$

Suppose, for example, that the roots are real and different (that is, $(\gamma - \beta)^2 + 4\alpha\delta > 0$); denote them by r_1 and r_2. Then we obtain

$$z(t) = \frac{(\gamma - r_1)z(0) - \alpha y(0)}{r_2 - r_1} e^{r_1 t} + \frac{(r_2 - \gamma)z(0) + \alpha y(0)}{r_2 - r_1} e^{r_2 t}, \qquad (15.25)$$

$$y(t) = \frac{(\beta - r_1)y(0) - \delta z(0)}{r_2 - r_1} e^{r_1 t} + \frac{(r_2 - \beta)y(0) + \delta z(0)}{r_2 - r_1} e^{r_2 t}. \qquad (15.26)$$

If r_1 and r_2 are both negative, then $z(t)$ and $y(t) \to 0$ as $t \to \infty$; the origin is then a limit point of the trajectory and is called a *node*. It is said to be a *stable* node, since all trajectories approach it (conversely, the case $r_1 > 0, r_2 > 0$ would correspond to an *unstable* node, since all trajectories then leave the origin). We notice also that, if $r_2 < r_1 < 0$, then as $t \to \infty$, we have

$$\frac{z(t)}{y(t)} \to \frac{(\gamma - r_1)z(0) - \alpha y(0)}{(\beta - r_1)y(0) - \delta z(0)}, \qquad (15.27)$$

but

$$\frac{\gamma - r_1}{-\delta} = \frac{-\alpha}{\beta - r_1}$$

by the definition of r_1, so that the ratio (15.27) is independent of $z(0)$ and $y(0)$ and has the value $(r_1 - \gamma)/\delta$. Thus, the trajectories become tangent to a common direction at the origin. [One exception has to be noted: if $z(0)$ and $y(0)$ satisfy $(\gamma - r_1)z(0) - \alpha y(0) = 0$, the first term in each of Equations (15.25) and (15.26) vanishes; the corresponding trajectory then meets the origin at a slope of $(r_2 - \gamma)/\delta$.]

15.5 Problems

15.5.1 Describe the behavior of $z(t)$ and $y(t)$, as given by Equations (15.25) and (15.26), for cases in which the roots of Equation (15.24) are real but of different sign, real and equal, complex, or purely imaginary. Show that the behavior of the phase plane trajectories near the origin is appropriately described, in each of these three cases, by the statement that the origin is a *saddle point*, a *node* (Are there any differences from the kind of node corresponding to $r_2 < r_1 < 0$?), a *spiral point*, or a *vortex point*, respectively.

15.5.2 In the light of the preceding Section and Problem 15.5.1, discuss the character of the singular points of Equations (15.1) and (15.2), Equation (15.11), and Problems 15.3.1, 15.3.2.

15.5.3 Analyze the oscillatory motion of a violin string across which a bow is being drawn. What happens if the rosin on the bow is not uniformly distributed?

15.5.4 A trajectory in the phase plane having the form of a simple closed curve is termed a *limit cycle;* it corresponds to periodic motion. Suppose that, for the system (15.16), (15.17), F and G are such that $\partial F/\partial z + \partial G/\partial y \neq 0$ within a certain region R; show (Bendixson) that no limit cycle can exist within R.

[HINT: Use the fact that, if C is any simple closed curve,

$$\int_C (F\,dy - G\,dz) = -\int_A \left(\frac{\partial F}{\partial z} + \frac{\partial G}{\partial y} \right) dA,$$

where the first integral is a counterclockwise contour integral, and where the second integral is over the area A enclosed by C.]

15.5.5 Suppose that a *Liapunov function* $V(y, z)$ exists in a circular region R surrounding the origin, with the property that $V(y, z)$ is nonnegative in R, and vanishes only at the origin; moreover,

$$\dot{V} = \frac{\partial V}{\partial y} G + \frac{\partial V}{\partial z} F \leq 0$$

in R, where F and G are the functions of Equations (15.16) and (15.17). Show that the origin is a stable point, in the sense that trajectories starting near the origin will not depart too far from the origin. Can you strengthen this result for the case $\dot{V} < 0$ for all y, z in R, except that $\dot{V}(0, 0) = 0$?

The phase plane discussion in Sections (15.2) and (15.3) was restricted to the *autonomous* case, in which the independent variable (t in Equations (15.20) and (15.21)) does not explicitly occur. Also, only the case of two dependent variables was considered.

For a more general discussion, see J. J. Stoker, *Nonlinear Vibrations*, Interscience, 1950, or L. A. Pars, Ibid., or M. Tabor, *Chaos and Integrability in Nonlinear Dynamics*, Wiley, 1989, or J. Guckenheimer and P. Holmes, *Nonlinear Oscillations*, Springer Verlag, 1983.

More on Difference Equations | 16

16.1 Second-Order Equations

In Equation (2.15), we obtained the solution of the general first-order difference equation. For convenience, we repeat that result here, in a slightly different form. If

$$F_{n+1} - a_n F_n = c_n, \qquad n = 0, 1, 2, \ldots, \tag{16.1}$$

where the a_n and b_n are prescribed functions of n, then it follows that

$$F_n = \alpha_n \left[F_0 + \frac{c_0}{\alpha_1} + \frac{c_1}{\alpha_2} + \cdots + \frac{c_{n-1}}{\alpha_n} \right], \tag{16.2}$$

where

$$\alpha_n = a_0 a_1 \cdots a_{n-1}.$$

Equation (16.1) is said to be of first order because values of F_j at only two adjoining values of j are involved, or alternatively, because Equation (16.1) involves F_n and the "first difference" $F_{n+1} - F_n$. Still another reason for this nomenclature is that, as we saw in the discussion of Equation (2.1), a first-order differential equation can be approximated by an equation of the form (16.1). For analogous reasons, the equation

$$F_{n+1} + a_n F_n + b_n F_{n-1} = c_n, \qquad n = 1, 2, 3, \cdots, \tag{16.3}$$

where a_n, b_n, c_n are prescribed functions of n, is termed a linear second-order difference equation, and we now turn to a discussion of its properties.

We begin, however, with the simpler case of a linear homogeneous equation with constant coefficients. Consider

$$F_{n+1} + aF_n + bF_{n-1} = 0, \quad n = 1, 2, 3, \ldots, \tag{16.4}$$

where a and b are constants. If F_0 and F_1 are prescribed, then Equation (16.4) determines F_2, F_3, ..., recursively, and of course uniquely. (Thus, if we can find any solution of Equation (17.4) involving F_0 and F_1 as parameters, that must be the most general solution.)

The reader can verify readily that Equation (16.4) has particular solutions of the form

$$F_n = r^n. \tag{16.5}$$

The constant r must satisfy the condition

$$r + a + \frac{b}{r} = 0,$$

which in general has the two solutions r_1 and r_2 given by

$$r_1 = \frac{-a + \sqrt{a^2 - 4b}}{2}, \quad r_2 = \frac{-a - \sqrt{a^2 - 4b}}{2}. \tag{16.6}$$

We are thus led to the more general solution

$$F_n = Ar_1^n + Br_2^n, \tag{16.7}$$

where A and B are arbitrary constants. If the equations

$$F_0 = A + B, \quad F_1 = Ar_1 + Br_2 \tag{16.8}$$

can be solved for A and B in terms of F_0 and F_1, then insertion of the resulting values of A and B into Equation (16.7) gives the necessarily unique solution of Equation (16.4) as

$$F_n = F_1 \frac{r_1^n - r_2^n}{r_1 - r_2} - F_0 r_1 r_2 \frac{r_1^{n-1} - r_2^{n-1}}{r_1 - r_2}. \tag{16.9}$$

Although Equations (16.8) do not yield values for A and B in the case in which $r_1 = r_2$ (i.e., $a^2 = 4b$), we can obtain the solution for this case by considering the limiting value of the right-hand side of Equation (16.9) as $r_2 \to r_1 = r$, say. Using L'Hospital's rule, we obtain for this limiting case

$$F_n = nF_1 r^{n-1} - (n-1)F_0 r^n. \tag{16.10}$$

Thus, the most general solution of Equation (16.4) has the form (16.9) if $r_1 \neq r_2$, and the form (17.10) if $r_1 = r_2 = r$.

If, however, $a^2 < 4b$, so that r_1 and r_2 are complex, it may be convenient to use an alternative form of Equation (16.9). We can write

$$F_n = R^n[C \cos n\theta + D \sin n\theta]$$

and determine R and θ (both real) so that this expression satisfies Equation (16.4). The reader should show that the result corresponding to Equation (16.9) is now

$$F_n = \frac{b^{n/2}}{\sin \theta}\left[\frac{F_1}{\sqrt{b}} \sin n\theta - F_0 \sin(n-1)\theta\right], \qquad (16.11)$$

where $\cos \theta = -a/(2\sqrt{b})$, with $0 < \theta < \pi$.

16.2 Problems

16.2.1 Solve the following equations:

(a)
$$F_{n+1} + 2F_n = F_{n-1}, \qquad F_0 = F_1 = 1;$$

(b)
$$F_{n+1} + F_n + F_{n-1} = 1, \qquad F_0 = 0, \quad F_1 = 1;$$

(c)
$$F_{n+1} + 2F_n + F_{n-1} = 0, \qquad F_0 = 1, \quad F_1 = 2;$$

(d)
$$F_{n+1} + F_n + F_{n-1} = 0, \qquad F_0 = 0, \quad F_{10} = 0.$$

16.2.2 The form of Equation (16.11) suggests that, for the case in which r_1 and r_2 are real, it should be possible to write Equation (16.9) in terms of hyperbolic functions. Do so.

16.2.3 Let a particle be constrained to move along the segment $(0, N)$ of the x-axis, where N is a positive integer. The particle is permitted to occupy only the integral positions $x = 0, 1, 2, \ldots, N$; at time 0 the particle is at position n, and at each subsequent time $1, 2, 3, \ldots$, it moves a unit distance in either the positive direction (with probability p) or the negative direction (with probability $1 - p$). The "barriers" at 0 and N are absorbent, so that the particle terminates its motion if it ever meets one of these barriers. If the probability that a particle (starting at $x = n$) is eventually absorbed by the barrier at N be denoted by P_n, show that the difference equation satisfied by P_n is

$$P_n = pP_{n+1} + (1-p)P_{n-1} \qquad (16.12)$$

(for which values of n?) and solve this equation. Consider also the special case $p = \frac{1}{2}$. What is the expected duration of the process?

(This random walk problem is equivalent to the famous "gambler's ruin" problem, in which one player (who wins each game with probability p) starts with n monetary units, the other with $N - n$ units; the game is continued until one player is bankrupt.)

16.2.4 If N and n are large, then one might anticipate that a plot of P_n versus n could be fitted by a rather smooth curve, denoted say by $y(x)$ (with $y(n) = P_n$). To a good approximation, we could also expect that, for example,

$$P_{n+1} = y(n+1) = y(n) + y'(n)\cdot(1) + \tfrac{1}{2}y''(n)\cdot(1)^2 + \cdots.$$

Use this device to replace Equation (16.12) by an ordinary differential equation, and solve that equation subject to the boundary conditions $y(0) = 0$, $y(N) = 1$. For what values of p, n, N is this approximation good?

16.2.5 Consider the set of difference equations

$$\begin{aligned}F_1 - a_0 F_0 &= c_0,\\ F_2 - a_1 F_1 &= c_1,\\ F_3 - a_2 F_2 &= c_2,\\ &\vdots\end{aligned}$$

To illustrate a technique which is sometimes useful, let us choose a set of multipliers, λ_n, in some manner to be determined, and then multiply the nth equation by λ_n. If the resulting equations are added, we see that a useful choice for the λ_n is given by $\lambda_{n-1} = a_{n-1}\lambda_n$ (why?). Carry out the details, so as to obtain again Equation (16.2). Can you devise a similar procedure for second-order equations?

16.3 The General Linear Equation

Consider now

$$F_{n+1} + a_n F_n + b_n F_{n-1} = c_n, \qquad n = 1, 2, 3, \ldots, \qquad (16.13)$$

where a_n, b_n, c_n **are functions of n. Again, if F_0 and F_1 are prescribed, Equation (16.13) determines all succeeding F_n uniquely, so that any solution containing F_0 and F_1 as parameters, whose values may be freely chosen, must be unique.**

Apart from the device mentioned in Problem (2.1.5), there is no general method whereby one can construct a formal solution for an arbitrary second-order difference equation. However, just as in the case of the second-order differential equation, it is enough to be able to find one solution of the homogeneous equation. For let G_n be such a known solution; i.e., let

$$G_{n+1} + a_n G_n + b_n G_{n-1} = 0, \qquad n = 1, 2, 3, \ldots . \qquad (16.14)$$

Multiply Equation (16.13) by G_n, Equation (16.14) by F_n, and subtract, to obtain

$$(F_{n+1} G_n - G_{n+1} F_n) - b_n(F_n G_{n-1} - G_n F_{n-1}) = c_n G_n. \qquad (16.15)$$

But this is now a first-order equation for the quantity Y_n defined by

$$Y_n = F_n G_{n-1} - G_n F_{n-1}, \qquad (16.16)$$

and its solution may be obtained by use of Equation (16.2). In turn, Equation (16.16) is then a first-order equation for F_n, whose solution is immediate [divide Equation (16.16) by $G_{n-1} G_n$]:

$$F_n = G_n \left[\frac{F_0}{G_0} + \frac{Y_1}{G_1 G_0} + \frac{Y_2}{G_2 G_1} + \cdots + \frac{Y_n}{G_n G_{n-1}} \right]. \qquad (16.17)$$

Since $Y_1 = F_1 G_0 - G_1 F_0$, the function Y_n obtained by solving Equation (16.15) will contain the parameters F_1 and F_0; thus, a specification of F_1 and F_0 is enough to completely determine the solution (16.17).

As a special case, let $c_n \equiv 0$; Equation (16.17) then provides the general solution for this homogeneous equation:

$$F_n = G_n \left[\frac{F_0}{G_0} + (F_1 G_0 - G_1 F_0) \right.$$

$$\left. \times \left\{ \frac{1}{G_1 G_0} + \frac{b_1}{G_2 G_1} + \frac{b_2 b_1}{G_3 G_2} + \cdots + \frac{b_{n-1} \cdots b_2 b_1}{G_n G_{n-1}} \right\} \right]. \qquad (16.18)$$

The quantity Y_n defined by Equation (16.16) is analogous to the Wronskian of an ordinary differential equation. In fact, if S_n and T_n are two solutions of the homogeneous equation (16.14), it follows from Equation (16.15) that

$$S_{n+1} T_n - T_{n+1} S_n = (b_1 b_2 \cdots b_n)(S_1 T_0 - T_1 S_0), \qquad (16.19)$$

so that, just as in the case of differential equations, $S_{n+1} T_n - T_{n+1} S_n$ cannot vanish for one value of n unless it vanishes for all values of n (we assume that $b_n \neq 0$, all n). If $S_{n+1} T_n - T_{n+1} S_n$ is nonzero, S_n and

T_n are said to be linearly independent; the most general solution of the homogeneous equation then has the form

$$AS_n + BT_n,$$

where A and B are arbitrary constants.

16.4 Problems

16.4.1 Construct a second-order linear homogeneous difference equation whose independent solutions are n and $n!$.

16.4.2 Solve the following equations:

(a)
$$F_{n+1} + (n + 1)F_n - 2n(n + 1)F_{n-1} = 0,$$

(b)
$$(1 + n)F_{n+1} - nF_n - F_{n-1} = 1,$$

(c)
$$(2n^2 + n - 1)F_{n+1} - 4n^2 F_n + (2n^2 - n - 1)F_{n-1} = 3,$$

(d)
$$(n + 1)F_{n+1} + nF_n - F_{n-1} = 0.$$

16.4.3 Two identical decks of 52 playing cards, each arranged in random order, are placed face down on a table. One card from each deck is drawn, and the two cards compared. The process is repeated until all cards are gone. What is the probability that at least one pair of cards consists of identical cards? (Let P_n be the probability that there is no such match, for the case in which each deck consists of n cards; find a difference equation for P_n.)

16.4.4 A difference equation, corresponding to the differential equation $x^2 y'' + xy' - y = 0$, for x in $(0, 1)$ is

$$\left(j^2 + \frac{j}{2}\right) y_{j+1} - (2j^2 + 1) y_j + \left(j^2 - \frac{j}{2}\right) y_{j-1} = 0.$$

Here $y_j = y(x_j) = y(\delta \cdot x)$, where δ is the spacing between mesh points. Determine its general solution, and compare its behavior with that of the solution to the differential equation.

16.5 Generating Functions

To illustrate the *method of generating functions*, consider the equation

$$(n + 1)F_{n+1} + nF_n + F_{n-1} = 0, \quad n = 1, 2, 3, \ldots. \quad (16.20)$$

Let z be a variable, and consider the power series

$$\phi(z) = \sum_{n=0}^{\infty} F_n z^n, \quad (16.21)$$

where we hope that this series converges for sufficiently small values of z. If we can find $\phi(z)$, then we can in turn determine F_n as the coefficient of z^n in the series expansion of ϕ.

Multiply Equation (16.20) by z^n and form the sum

$$\sum_{n=1}^{\infty} [(n + 1)F_{n+1}z^n + nF_n z^n + F_{n-1}z^n] = 0. \quad (16.22)$$

To evaluate the first term, we write

$$\sum_{n=1}^{\infty} (n + 1)F_{n+1}z^n = \frac{d}{dz}\sum_{n=1}^{\infty} F_{n+1}z^{n+1} = \frac{d}{dz}[\phi - F_1 z - F_0].$$

Treating the other terms similarly, Equation (16.21) becomes a first-order differential equation for ϕ:

$$[\phi - F_1 z - F_0]' + z[\phi - F_0]' + z\phi = 0. \quad (16.23)$$

At $z = 0$, we have $\phi = F_0$; the solution of Equation (16.23), subject to this initial condition, is

$$\phi = F_0 e^{-z}(1 + z) + F_1 e^{-z}(1 + z)\int_0^z \frac{e^{\xi}}{(1 + \xi)^2}\,d\xi$$

$$= (F_0 + F_1)e^{-z}(1 + z) - F_1 + F_1 e^{-z}(1 + z)\int_0^z \frac{e^{\xi}}{1 + \xi}\,d\xi. \quad (16.24)$$

Thus, F_n is the coefficient of z^n in the series expansion of the right hand side of Equation (16.24). It may, of course, be tedious to obtain such a series expansion for the integral term in Equation (16.24); we would have to expand the integrand, and integrate term by term (although the labor can be reduced by the exercise of some ingenuity). Alternatively, we could recognize that the integral involves something called an incomplete Gamma function, and could look up its properties (e.g., in *Hand-*

book of *Mathematical Functions*, National Bureau of Standards, Applied Mathematics Series 55, 1964) in order to obtain an elegant expression for F_n. However, we can save ourselves some work here by noting that if we set $F_1 = 0$ and $F_0 = 1$ in Equation (16.24), we obtain a particular solution of Equation (16.20) as the coefficient of z^n in the expansion of $(1 + z)e^{-z}$:

$$G_n = (-1)^n \left[\frac{1}{(n-1)!} - \frac{1}{n!} \right], \quad n = 1, 2, 3, \ldots. \quad (16.25)$$

The method of the last section can then be used (and should be used by the reader) to obtain the general solution of Equation (16.20) in reasonably compact form.

As a second example, we choose one which involves boundary conditions at the two ends of an interval. Consider again the difference equation (16.12):

$$P_n = pP_{n+1} + (1-p)P_{n-1} \quad (16.26)$$

for $n = 1, 2, \ldots, N-1$, where $P_0 = 0$ and $P_N = 1$. We now form the generating function

$$\psi(z) = \sum_{n=0}^{N} P_n z^n, \quad (16.27)$$

which is a polynomial rather than an infinite series as in the preceding example. Multiply Equation (16.26) by z^n and sum over n from 1 to $N-1$ to obtain

$$\psi(z)[z - p - (1-p)z^2]$$
$$= z^{N+1} - (1-p)z^{N+2} - pP_1 z - (1-p)P_{N-1} z^{N+1}, \quad (16.28)$$

where we have used $P_0(0)$ and $P_N = 1$. Equation (16.28) apparently contains two unknown constants, namely, P_1 and P_{N-1}; how can we determine them?

The trick is to observe that the left hand side of Equation (16.28) vanishes for $z = 1$ and for $z = p/(1-p)$; the right hand side must then also vanish for these two values of z. This gives two linear equations for P_1 and P_{N-1}:

$$pP_1 + (1-p)P_{N-1} = p,$$

$$\frac{p^2}{1-p} P_1 + (1+p)\left(\frac{p}{1-p}\right)^{N+1} P_{N-1}$$
$$= \left(\frac{p}{1-p}\right)^{N+1} - (1-p)\left(\frac{p}{1-p}\right)^{N+2}. \quad (16.29)$$

Once Equations (16.29) have been solved for P_1 and P_{N-1}, $\psi(z)$ is completely determined by Equation (16.28), and P_n for any n is then obtained as the coefficient of z_n in $\psi(z)$. The reader should complete the details of this process and verify that his result agrees with that of Problem (16.2.3).

16.6 Problems

16.6.1 Use the generating function method to solve Equation (16.4), and also those equations encountered in Problems (16.4.2) and (16.4.3).

16.6.2 The Legendre polynomials $P_n(z)$ ($n = 0, 1, 2, \ldots$) satisfy the difference equation

$$(n + 1)P_{n+1}(z) - (2n + 1)zP_n(z) + nP_{n-1}(z) = 0.$$

Moreover, $P_0(z) = 1$, $P_1(z) = z$. Use the generating function method to solve this equation, so as to deduce that

$$\frac{1}{\sqrt{1 - 2z\xi + \xi^2}} = P_0(z) + \xi P_1(z) + \xi^2 P_2(z) + \cdots. \quad (16.30)$$

16.6.3 The Bessel functions of the first kind, of integral order, satisfy the recurrence relation

$$\frac{2n}{z} J_n(z) = J_{n-1}(z) + J_{n+1}(z), \quad n = \ldots, -2, -1, 0, 1, 2, \ldots. \quad (16.31)$$

Defining the generating function

$$\phi(z, \xi) = \sum_{n=-\infty}^{\infty} J_n(z)\xi^n,$$

solve Equation (16.31) by the generating-function method to determine $\phi(z, \xi)$. Once ϕ has been found, what properties of the $J_n(z)$ can be deduced from it?

16.6.4 Bessel's differential equation for functions of order zero is $xy'' + y' + xy = 0$. If we use a mesh-point spacing of δ and define $x_j = j\delta$ and $y_j \simeq y(x_j)$, a difference equation approximation is

$$(j + \tfrac{1}{2})y_{j+1} + (\delta^2 - 2)jy_j + (j - \tfrac{1}{2})y_{j-1} = 0 \quad (16.32)$$

Use the generating function method to solve Equation (16.32), if the value y_0 is specified, and compare the solution with $J_0(x)$.

16.7 Nonlinear Difference Equations

An example of a nonlinear difference equation problem is encountered in the following iterative process. Let α be a prescribed positive constant, and define

$$w_0 = \text{any number},$$

$$w_1 = 1 + \frac{\alpha}{w_0},$$

$$w_2 = 1 + \frac{\alpha}{w_1} = 1 + \frac{\alpha}{1 + (\alpha/w_0)}, \qquad (16.33)$$

$$w_3 = 1 + \frac{\alpha}{w_2} = 1 + \left[\alpha \bigg/ \left\{1 + \frac{\alpha}{1 + (\alpha/w_0)}\right\}\right],$$

$$\vdots$$

If this process converges to a limit w_∞, we write

$$w_\infty = 1 + \cfrac{\alpha}{1 + \cfrac{\alpha}{1 + \cfrac{\alpha}{1 + \cfrac{\alpha}{1 + \cfrac{\alpha}{\vdots}}}}} \qquad (16.34)$$

and describe the expression on the right hand side of Equation (16.34) as a *continued fraction*. To find w_∞ (if it exists) is easy, since Equation (16.34) implies that

$$w_\infty = 1 + \frac{\alpha}{w_\infty},$$

which has the solution $w_\infty = \frac{1}{2}(1 + \sqrt{1 + 4\alpha})$. To find w_n for n finite is less easy; we proceed as follows.

The general difference equation describing the process (16.33) is

$$w_{n+1} = 1 + \frac{\alpha}{w_n}$$

or, alternatively,

$$w_{n+1}w_n - w_n = \alpha. \qquad (16.35)$$

We alter this to a homogeneous equation, by defining

$$w_n = y_n + \beta, \tag{16.36}$$

where the constant β satisfies the equation

$$\beta^2 - \beta = \alpha.$$

For definiteness, we choose $\beta = \frac{1}{2}(1 - \sqrt{1 + 4\alpha})$. Equation (16.35) becomes

$$y_{n+1}y_n + \beta y_{n+1} + (\beta - 1)y_n = 0. \tag{16.37}$$

Dividing through by $y_n y_{n+1}$ and setting

$$\xi_n = \frac{1}{y_n}, \tag{16.38}$$

we obtain

$$1 + \beta \xi_n + (\beta - 1)\xi_{n+1} = 0, \tag{16.39}$$

which has the form of Equation (16.1). Actually, we can write down its solution by inspection as

$$\xi_n = \frac{1}{1 - 2\beta} + \left(\frac{\beta}{1 - \beta}\right)^n \left(\xi_0 - \frac{1}{1 - 2\beta}\right). \tag{16.40}$$

Inserting $\xi_0 = 1/y_0 = 1/(w_0 - \beta)$ into Equation (16.40) and expressing ξ_n in terms of w_n, we obtain

$$w_n = \frac{1}{2}(1 - \sqrt{1 + 4\alpha}) + \left[\frac{1}{\sqrt{1 + 4\alpha}} + \left(\frac{1 - \sqrt{1 + 4\alpha}}{1 + \sqrt{1 + 4\alpha}}\right)^n \right.$$
$$\left. \times \left\{\frac{1}{w_0 - \frac{1}{2}(1 - \sqrt{1 + 4\alpha})} - \frac{1}{\sqrt{1 + 4\alpha}}\right\}\right]^{-1}. \tag{16.41}$$

Thus, the continued fraction process will converge, for any $w_0 \neq \frac{1}{2}(1 - \sqrt{1 + 4\alpha})$, to $w_\infty = \frac{1}{2}(1 + \sqrt{1 + 4\alpha})$.

The reader should note that any nonlinear equation of the form

$$F_{n+1}F_n + aF_{n+1} + bF_n + c = 0$$

may be solved by essentially the same process as that used to solve Equation (16.35).

As a second example, let $Q > 0$ be prescribed. A well-known computational procedure for determining $\sqrt{Q}$ is to start with some initial approximation y_0 and to compute recursively

$$y_{n+1} = \frac{1}{2}\left(y_n + \frac{Q}{y_n}\right), \qquad n = 0, 1, 2, 3, \ldots. \qquad (16.42)$$

This process converges rapidly, for if we define the relative error at the nth step, ϵ_n, by $y_n = \sqrt{Q}(1 + \epsilon_n)$, then Equation (16.42) leads to

$$\epsilon_{n+1} = \frac{1}{2}\epsilon_n^2 + 0(\epsilon_n^3). \qquad (16.43)$$

To solve Equation (16.42) explicitly, we carry out the following sequence of transformations:

$$y_n = \sqrt{Q}(w_n + 1), \qquad w_n = \frac{1}{z_n}, \qquad z_n = x_n - \frac{1}{2}. \qquad (16.44)$$

We thus obtain

$$x_{n+1} = 2x_n^2,$$

so that

$$x_n = 2^{(2^n-1)}x_0^{(2^n)}.$$

In terms of y_n, we have

$$y_n = \sqrt{Q} + \left[2\sqrt{Q} \Big/ \left\{\left(1 + \frac{2\sqrt{Q}}{y_0 - \sqrt{Q}}\right)^{(2^n)} - 1\right\}\right]. \qquad (16.45)$$

Note that Equation (16.45) is compatible with Equation (16.43) and that, as $n \to \infty$, then $y_n \to \sqrt{Q}$ or $-\sqrt{Q}$ according to whether $y_0 > 0$ or $y_0 < 0$, respectively.

16.8 Problems

16.8.1 Show that Equation (16.42) is equivalent to the use of the Newton-Raphson method (also called the *tangent* method) of elementary calculus, as applied to the problem of determining where the function $y^2 - Q$ vanishes. Derive a similar iterative process for taking cube roots, investigate its rate of convergence, and discuss the resulting difference equation.

16.8.2 If the *secant* method†, rather than the *tangent* method, is used to find the value x^* of x at which a function $f(x)$ vanishes, the sequence of approximations to x^* can be described by the difference equation

---

† A discussion of zero-finding methods will be found in C. Pearson, *Numerical Methods in Engineering and Science*, Van Nostrand, 1986.

$$x_{n+2} = x_{n+1} - \frac{(x_{n+1} - x_n)f(x_{n+1})}{f(x_{n+1}) - f(x_n)}. \tag{16.46}$$

Write $x_n = x^* + \epsilon_n$, assume f' and f'' to exist, and show that

$$\epsilon_{n+2} \sim c\epsilon_{n+1}\epsilon_n$$

as $n \to \infty$. Determine c. Show also that

$$\epsilon_n \sim k\epsilon_{n-1}^{1.62}$$

and determine k.

16.8.3 Discuss the continued fraction

$$w = 1 + \cfrac{\alpha}{1 - \cfrac{\alpha}{1 + \cfrac{\alpha}{1 - \cfrac{\alpha}{\vdots}}}}$$

16.8.4 An electric circuit is constructed by adding sections as shown in Figure 16.1. Each resistor is one ohm.

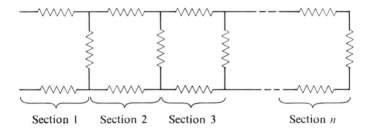

Figure 16.1 Electric circuit with n sections.

Show that the input resistance corresponding to the use of n sections, R_n, satisfies the difference equation

$$R_{n+1} = 2 + \frac{R_n}{1 + R_n},$$

and solve this equation. Modify your analysis so as to apply to the cases (a) each resistance in the nth section has a value of $(\tfrac{1}{2})^n$ ohms, (b) each resistance in the nth section has a value of $1/n$ ohms.

16.8.5 A sophisticated gambler is offered the following proposition. "On the table before you are twelve pieces of paper, each bearing a certain number on its down face. The numbers can be anything at all, except that they are all different. You may pick up these pieces of paper, one at a time; as you pick up each piece and look at its number, you may decide to either stop with that piece or else discard it and proceed. If you manage to stop with that piece of paper whose number is highest among all twelve, you win; otherwise you lose." After some thought, the gambler decides that, given n pieces of paper carrying n different numbers, his optimal strategy is to discard the first k of these pieces of paper, recording the largest number, M, written on any of them. He then proceeds and stops with the next piece bearing a number larger than M, if any. Find a difference equation satisfied by $P(k, n)$ (the probability of winning by the use of this strategy), solve it, and determine the optimal value of k.

16.9 Further Properties of Difference Equations

The following exercises illustrate a number of properties of difference equations which are analogous to the corresponding properties of differential equations. Included are problems on higher-order difference equations, on Green's functions, and on eigenvalues and eigenfunction expansions.

16.9.1 (a) Obtain a second-order difference equation for y_n, if

$$y_{n+1} + a_n y_n + b_n z_n = c_n, \qquad z_{n+1} + \alpha_n y_n + \beta_n z_n = \gamma_n, \qquad (16.47)$$

where a_n, b_n, c_n, α_n, β_n, γ_n are given functions of n. Can this always be done?

(b) Write the general linear pth-order difference equation for the functions y_n in terms of a set of p first-order equations involving p functions y_n, z_n, ...

16.9.2 In any one toss of a particular weighted coin, the probability of "heads" is p. Show that the probability of a run of at least h heads in succession, in a total of n tosses (with $n > h$), can be determined from the solution of a difference equation of order $(h + 1)$, and solve this equation.

16.9.3 Show that the general second-order linear difference equation (16.13) can be written in the form
$$p_{n+1}(F_{n+1} - F_n) - p_n(F_n - F_{n-1}) + q_n F_n = s_n \qquad (16.48)$$
by appropriate choices for p_n, q_n, and s_n.

16.9.4 Define the difference operator L_n by
$$L_n(H_n) = p_{n+1}(H_{n+1} - H_n) - p_n(H_n - H_{n-1}) + q_n H_n, \qquad (16.49)$$
where p_n, q_n are given functions of n. Show that
$$\sum_{n=1}^{N-1} [H_n L_n(G_n) - G_n L_n(H_n)]$$
$$= p_N(G_N H_{N-1} - G_{N-1} H_N) - p_1(G_1 H_0 - G_0 H_1). \qquad (16.50)$$

To what does this identity correspond in the theory of differential equations?

16.9.5 Define Green's functions, $G_n^{(r)}$, for Equation (16.48) so as to satisfy
$$p_{n+1}(G_{n+1}^{(r)} - G_n^{(r)}) - p_n(G_n^{(r)} - G_{n-1}^{(r)}) + q_n G_n^{(r)} = \delta_n^r, \qquad (16.51)$$
where $\delta_n^r = 1$ for $r = n$, and $\delta_n^r = 0$ for $r \neq 0$. Let $G_0^{(r)} = G_N^{(r)} = 0$.
Show that the following equation holds:
$$F_r = \sum_{n=1}^{N-1} G_n^{(r)} s_n - G_1 p_1 F_0 - G_{N-1} p_N F_N. \qquad (16.52)$$

Show also that $G_n^{(r)}$ is symmetric in the two indices r and n.

16.9.6 Consider a number of uniformly spaced masses, each of mass m, as shown in Figure 16.2.

Figure 16.2 Springs and masses in series.

Let the spring constant of each spring be k. If y_i denotes the horizontal displacement of the ith mass, the equation of motion is
$$m \frac{d^2 y_i}{dt^2} = k(y_{i+1} - y_i) - k(y_i - y_{i-1}). \qquad (16.53)$$

As end conditions, we take $y_0 = y_N = 0$. A "normal mode" kind of oscillatory motion is one for which all displacements are exactly in phase (or exactly out of phase); to obtain such a mode, we set $y_i = A_i \sin \omega t$ in Equation (16.53) to give

$$A_{i+1} - 2A_i + A_{i-1} + \lambda A_i = 0, \tag{16.54}$$

where $\lambda = m\omega^2/k$, $A_0 = A_N = 0$. Equation (16.54) now involves an eigenvalue λ; only for certain values of λ can a non-trivial solution satisfying the given end conditions be found. (a) Use the generating function method to determine, in as simple a form as you can, all eigenvalues and eigenfunctions of Equation (16.54). (b) Let N be large; replace Equation (16.54) by a differential equation and solve it so as to obtain approximations for certain modes. Make appropriate comments.

16.9.7 Consider the eigenvalue problem

$$p_{n+1}(F_{n+1} - F_n) - p_n(F_n - F_{n-1}) + q_n F_n + \lambda r_n F_n = 0 \tag{16.55}$$

with $F_0 = F_N = 0$, where p_n, q_n, r_n are given (real) functions of n. If $r_n > 0$ for all n, show that all eigenvalues are real, and that two eigenfunctions $F_n^{(1)}$ and $F_n^{(2)}$ corresponding to different eigenvalues $\lambda^{(1)}$ and $\lambda^{(2)}$ satisfy the orthogonality condition

$$\sum_{n=1}^{N-1} r_n F_n^{(1)} F_n^{(2)} = 0. \tag{16.56}$$

Discuss the possibility of the expansion of a given function, M_n, defined for $n = 1, 2, \ldots, N-1$, in terms of the $F_n^{(j)}$.

16.10.8 (a) Show that any set of $(N-1)$ quantities, $y_1, y_2, \ldots, y_{N-1}$ can be written as

$$y_j = \sum_{n=1}^{N-1} b_n \sin \frac{n\pi j}{N}, \qquad j = 1, 2, \ldots, (N-1),$$

where (16.57)

$$b_n = \frac{2}{N} \sum_{j=1}^{N-1} y_j \sin \frac{n\pi j}{N}, \qquad n = 1, 2, \ldots, (N-1).$$

(b) Find a somewhat similar formula involving cosines.

16.10.9 (a) Show that if all $p_n > 0$, all $q_n < 0$, and all $r_n > 0$, then all eigenvalues of Equation (16.55) are positive.

(b) Let the conditions of Part (a) hold. Is it true that

$$\lambda_{\min} \leq \frac{\sum_{n=1}^{N} p_n(W_n - W_{n-1})^2 - \sum_{n=1}^{N-1} q_n W_n^2}{\sum_{n=1}^{N-1} r_n W_n^2}, \qquad (16.58)$$

where the set $(W_0, W_1, \ldots, W_N)$ may be arbitrarily chosen? If not, obtain a correct inequality by appropriately modifying Equation (16.58).

Numerical Methods | 17

Although the techniques displayed in the preceding pages provide adequate tools for the resolution of many questions involving differential equations, there are a number of problems for which these tools are inadequate. Numerical methods, together with the use of computers, permit one to deal with many of these more recalcitrant questions. To illustrate the general idea, consider again Equations (15.1) and (15.2):

$$\frac{dP}{dt} = aPF - kP, \qquad (17.1)$$

$$\nu \frac{dF}{dt} = c - F - \beta P, \qquad (17.2)$$

where a, k, c, ν, β are constants. Let $P(0) = P_0$ and $F(0) = F_0$ be prescribed; we want to determine $P(t)$ and $F(t)$ for values of t satisfying $0 < t < T$, where T is some chosen upper limit.

In using a numerical procedure, we content ourselves with the (approximate) determination of $P(t)$ and $F(t)$ at a set of discrete values of t. Divide the whole interval T into a number N of sub-intervals, each of length $h = T/N$. Let j be an integer; we hope to determine $P(jh)$ and $F(jh)$ for each value of j in $0 \leq j \leq N$. At time $t_j = jh$, the value of the right-hand side of Equation (17.1) is

$$aP(jh) \cdot F(jh) - kP(jh),$$

and a natural approximation for the left-hand side is

$$\frac{P([j+1]h) - P(jh)}{h}.$$

Treating Equation (17.2) similarly, we are led to the set of difference equations

$$\frac{P_{j+1} - P_j}{h} = aP_jF_j - kP_j, \qquad (17.3)$$

$$v\frac{F_{j+1} - F_j}{h} = c - F_j - \beta P_j. \qquad (17.4)$$

Here we have introduced a new notation, writing P_j rather than $P(jh)$, not only for simplicity, but also to emphasize the fact that, since Equations (17.3) and (17.4) are not exactly equivalent to Equations (17.1) and (17.2), their solutions P_j and F_j will not coincide exactly with $P(jh)$ and $F(jh)$. Thus, P_j and F_j denote the computed approximations to $P(jh)$ and $F(jh)$ obtained by solving the set of difference equations (17.3) and (17.4).

The actual solution of Equations (17.3) and (17.4) on a digital computer is straightforward. Given P_0 and F_0, Equations (17.3) and (17.4), with $j = 0$ permit the determination of P_1 and F_1; with $j = 1$, they then permit the determination of P_2 and F_2, and so on, recursively. A simple repetitive process of this kind is easily programmed for a digital computer. In the present case, tens or even hundreds of thousands of values for P_j and F_j, each calculated to a dozen or so significant figures, will be generated each second by a modern computer.

We expect that Equations (17.3) and (17.4) will represent Equations (17.1) and (17.2) more and more accurately as h is taken smaller and smaller. In practice, we might solve Equations (17.3) and (17.4) twice on the computer (once with the chosen value of h, and once with a value of h only half as large) to see if the solution in the two cases is appreciably different; if not, the original value of h is probably small enough. Observe, however, that once a solution of Equations (17.3) and (17.4) has been found and tested for accuracy in this way, it still corresponds to only one particular choice for the constants a, k, v, c, β, P_0, F_0; a different choice for any of these constants will require a new computation.

Other boundary conditions might, of course, have been prescribed. For example, $F(0)$ and $F(T)$ might have been given; we could then start by guessing $P(0)$, carry out the computation as above, and then iteratively adjust $P(0)$ until F_N attained the desired value $F(T)$. Alternatively, $P(0)$ and $F'(0)$ could be prescribed; however, use of the given values for $F'(0)$ and $P(0)$ in Equation (17.2) then gives $F(0)$, so we can again start with known values for $P(0)$ and $F(0)$.

The numerical procedure described above is a natural one. However, a number of questions might be asked. Is the equation set (17.3), (17.4), the "best" approximation to Equations (17.1), (17.2)? Would it, for example, be more accurate to approximate $P'(t_j)$ by $[P_{j+1} - P_{j-1}]/(2h)$, and if so would we pick up any extraneous solutions corresponding to the fact that the resulting difference equations would each now be of second order? In testing the accuracy of a computed solution by repeating the calculation for a different value of h, how much of a change in the solution should be considered significant? A computer carries only a finite number of significant digits, and the fact that all subsequent digits are lost in each calculation leads to what is called "round-off error"; when will such round-off error be important? Suppose we are interested not so much in the solution functions as in their derivatives; how can we obtain adequate accuracy? To these and other questions we now turn our attention.

Some insight can be obtained by studying problems for which we can obtain an exact solution. An example is afforded by

$$\alpha^2 y'' + y = \sin x, \qquad \alpha > 0, \tag{17.5}$$

with $y(0) = y'(0) = 0$. The true solution is

$$y = \frac{\sin x - \alpha \sin (x/\alpha)}{1 - \alpha^2} \tag{17.6}$$

for $\alpha \neq 1$, and (by L'Hospital's rule applied to Equation (17.6))

$$y = \tfrac{1}{2}(\sin x - x \cos x) \tag{17.7}$$

for $\alpha = 1$. Using an x-interval of length h, a simple finite difference approximation to Equation (17.5) is

$$\alpha^2 \left[\frac{y_{j+1} - 2y_j + y_{j-1}}{h^2} \right] + y_j = \sin x_j, \tag{17.8}$$

where y_j is the approximation to $y(x_j)$ with $x_j = jh$ for $j = 0, 1, 2, \ldots$. The initial conditions for Equation (17.8), corresponding to $y(0) = y'(0) = 0$ for Equation (17.5), could reasonably be taken as $y_0 = y_1 = 0$. The exact solution of Equation (17.8), which would be obtained by a computer operating without round-off error, is easily found by the methods of Chapter 16. It is given by

$$y_j = \frac{\sin jh - (\sin h/\sin \theta) \sin j\theta}{1 - 4(\alpha^2/h^2) \sin^2 (h/2)} \qquad (17.9)$$

for $\sin h/2 \neq h/2\alpha$, where $\sin (\theta/2) = h/2\alpha$. For the special case $\sin (h/2) = h/2\alpha$, Equation (17.9) must be replaced by

$$y_j = \frac{\cos h \sin jh - j \sin h \cos jh}{2 \cos^2 (h/2)}. \qquad (17.10)$$

Let us first compare Equations (17.6) and (17.9). In the case where $h \ll 1$, then we know that $\sin h/2 \simeq h/2$, and the denominator of Equation (17.9) becomes approximately $1 - \alpha^2$. Moreover, the numerator term $\sin jh$ is the same as $\sin x_j$ in Equation (17.6), where $x_j = jh$. However, the second numerator term,

$$-\frac{\sin h}{\sin \theta} \sin j\theta = -\frac{\sin h}{\sin \theta} \sin \left(\frac{\theta}{h} x_j\right),$$

can be close to $-\alpha \sin (x_j/\alpha)$ [the second numerator term of Equation (17.6)] only if $h \ll \alpha$ also. But even if this be the case, the argument $\theta x_j/h$ of the sine term will differ from the argument x_j/α of the corresponding sine term in Equation (17.6) by

$$\frac{\theta}{h} x_j = x_j \frac{2 \sin^{-1} (h/2\alpha)}{h} = \frac{2x_j}{h} \left[\frac{h}{2\alpha} + \frac{1}{6}\left(\frac{h}{2\alpha}\right)^3 + \cdots\right]$$
$$= \frac{x_j}{\alpha}\left[1 + \frac{1}{6}\left(\frac{h}{2\alpha}\right)^2 + \cdots\right]. \qquad (17.11)$$

Thus there will be a substantial discrepancy between the results of Equations (17.6) and (17.9) when x_j is large enough for

$$\frac{x_j}{6\alpha}\left(\frac{h}{2\alpha}\right)^2 \qquad (17.12)$$

to be a significant part (say .1) of a radian. We conclude that the numerical process must be designed so that $h \ll 1$, $h \ll \alpha$, and that the results can only be used for values of x_j such that the expression (17.12) is less than about .1. Similar remarks apply for the exceptional cases (17.7) and (17.10); we observe, however, that these two exceptional cases do not arise for quite the same values of α. Thus the choice $\alpha = 1$ leads to an exact solution (17.7) whose magnitude of oscillation increases with x, but to a numerical solution (17.9) whose magnitude of oscillation is constant.

Even apart from this exceptional case, the parameter α plays an important role in the constraint it imposes on the size of h via the condition $h \ll a$. This corresponds, of course, to the condition that we use many sub-intervals within a wave-length of the complementary solutions to Equation (17.5), and this is a very natural requirement. However, if α is very small (say 10^{-6} or so) the condition $h \ll \alpha$ can become burdensome. In the foregoing, we have assumed that round-off error is negligible. In an actual computer, the fact that each calculation introduces a small error due to the dropping of all significant figures beyond, say, the 10th can result in a substantial cumulative error if many sub-intervals have been used. This situation may well arise for small values of α, and especially so if the differential operator has the form $\alpha^2 y'' - y$ rather than $\alpha^2 y'' + y$; for then the exponential-type solutions of the corresponding difference equation can magnify small round-off errors as the calculation proceeds. We will return to this point later.

17.1 Problems

17.1.1 Replace the differential equation of nth-order,

$$y^{(n)} = f(x, y, y', y'', \ldots y^{(n-1)}),$$

by a set of n first-order differential equations, involving the quantities $y, z_1 = y', z_2 = y'', \ldots, z_{n-1} = y^{(n-1)}$, and write down an approximating set of difference equations of the kind used in connection with Equations (17.1) and (17.2). Had this procedure been used for Equation (17.5), would the difference equation set have been equivalent to Equation (17.8)?

17.1.2 If $y(0) = y'(0) = 0$, then we can write

$$y(h) = y(0) + hy'(0) + \tfrac{1}{2}h^2 y''(0) + \tfrac{1}{6}h^3 y'''(0) + \cdots$$

$$= 0 + 0 + 0 + \frac{1}{6}\frac{h^3}{\alpha^2} + \cdots,$$

where we have made use of Equation (17.5). Thus, rather than set $y_0 = y_1 = 0$ in order to approximate the conditions $y(0) = y'(0) = 0$, we could set $y_0 = 0$, $y_1 = h^3/6\alpha^2$. Discuss the difference that this would have made in the solution of Equation (17.8).

17.1.3 Replace Equation (17.5) by the equation

$$\alpha^2 y'' - y = \sin x$$

and develop the corresponding discussion.

17.1.4 Let A be a positive constant, and consider the set of equations

$$y' = (-\cosh A)y + (\sinh A)z, \qquad z' = (\sinh A)y + (-\cosh A)z.$$

In solving this set of equations numerically, for instance over the interval $(0, 1)$, why must a surprisingly small mesh interval h be taken, even for moderate values of A, say 3 or so? Note that the equation set itself looks quite innocuous.

17.2 Discretization Error; Some One-Step Methods

Suppose we try to solve the problem

$$y' = f(x, y), \qquad y(a) = \text{prescribed} \tag{17.13}$$

for x in (a, b), where f is a given function, by means of the numerical procedure

$$\frac{y_{j+1} - y_j}{h} = f(x_j, y_j),$$

where $y_0 = y(a)$, $x_j = a + jh$ for $j = 0, 1, 2, \ldots$ and where y_j denotes the computed approximation to $y(x_j)$. We then repeatedly apply the equation

$$y_{j+1} = y_j + hf(x_j, y_j), \tag{17.14}$$

starting with $y_0 = y(a)$, and so generate values for $y_1, y_2, y_3, \ldots$ in succession. We can think of Equation (17.14) as a consequence of the approximation of y' by $(y_{j+1} - y_j)/h$, a consequence of the use of the first two terms of a Taylor series expansion for $y(x_{j+1})$ in terms of quantities evaluated at the point x_j, or as an approximation to the integral expression

$$y(x_{j+1}) - y(x_j) = \int_{x_j}^{x_{j+1}} f(x, y(x))\, dx.$$

The method of Equation (17.14) is conventionally associated with the name of Euler, although it is so natural an approach to the numerical solution of Equation (17.13) that the nomenclature hardly confers the honor intended.

To evaluate the error resulting from use of Equation (17.14), we ask to what extent the true solution $y(x)$ fails to satisfy Equation (17.14). Here a useful tool is Taylor's theorem, which states that, if the first $n + 1$ derivatives of $g(x)$ are continuous in $[x_j, x_{j+1}]$, then, for x in this interval,

$$g(x) = g(x_j) + \frac{g'(x_j)}{1!}(x - x_j) + \frac{g''(x_j)}{2!}(x - x_j)^2 + \cdots$$
$$+ \frac{g^{(n)}(x_j)}{n!}(x - x_j)^n + R_n(x), \quad (17.15)$$

where

$$R_n(x) = \frac{g^{(n+1)}(\xi)}{(n+1)!}(x - x_j)^{n+1}$$

with $x_j \leq \xi(x) \leq x$.

Thus, in particular, for the solution $y(x)$ of Equation (17.13) we have

$$y(x_{j+1}) = y(x_j) + hf(x_j, y(x_j)) + \tfrac{1}{2}h^2 y''(\xi), \quad (17.16)$$

which may be compared with Equation (17.14). Here $x_j \leq \xi \leq x_{j+1}$. The discrepancy between Equations (17.14) and (17.16) is termed the *discretization* (or *truncation*) error; in this case, it is said to be of order h^2.

This error occurs at each application of Equation (17.14). If we use this equation a total of n times in covering the interval $[a, b]$ with $n = (b - a)/h$, then to a rough order of magnitude we can expect the accumulated error to be of order nh^2, that is, of order h since n is of order $1/h$. Thus, once h is sufficiently small that Euler's method is reasonably accurate, halving h should halve the error. Similarly, if we find for a subsequent method that the discretization error is of order h^p, then the accumulated error over a fixed $[a, b]$ interval can be expected to be of order h^{p-1}. Of course, this argument is indicative only, since the calculational procedure might tend to greatly amplify errors; we will in fact meet such situations later on. Nevertheless, experience shows that this general expectation is frequently valid.

The Euler method is said to be a *one step* method, since it does not use calculated values of y_p for $p < j$ in order to compute y_{j+1}. A simple way in which to obtain a more accurate one step process is to add more terms to the partial Taylor series of Equation (17.14). Thus, since

$$y(x_{j+1}) = y(x_j) + hy'(x_j) + \tfrac{1}{2}h^2 y''(x_j) + \cdots \quad (17.17)$$

and since

$$y'(x) = f(x, y), \quad y''(x) = f_x(x, y) + f_y(x, y) \cdot y' = f_x + f_y f,$$

we are led to

$$y_{j+1} = y_j + hf(x_j, y_j) + \tfrac{1}{2}h^2[f_x(x_j, y_j) + f_y(x_j, y_j) \cdot f(x_j, y_j)] \quad (17.18)$$

§17.2] Discretization Error; Some One-Step Methods

as an improvement over Equation (17.14). The discretization error is now of order h^3. Further terms can be included, if desired, so as to obtain a still smaller discretization error; of course, one pays for this by performing more computations at each step.

As an example, consider the problem wherein

$$y' = x^2 - y^2, \quad y(0) = 1$$

(cf Equation (15.4)) and we want to find $y(1)$. We try three formulas, corresponding to the use of two, three, and four terms of the Taylor series, respectively. They are

$$y_{j+1} = y_j + h(x_j^2 - y_j^2), \tag{17.19}$$

$$y_{j+1} = y_j + h(x_j^2 - y_j^2) + h^2(x_j - y_j x_j^2 + y_j^3), \tag{17.20}$$

$$y_{j+1} = y_j + h(x_j^2 - y_j^2) + h^2(x_j - y_j x_j^2 + y_j^3)$$
$$+ \tfrac{1}{3}h^3(1 - 2x_j y_j - x_j^4 + 4x_j^2 y_j^2 - 3y_j^4). \tag{17.21}$$

TABLE 17.1 Computed Values of $y(1)$ for $y' = x^2 - y^2$, $y(0) = 1$

n	Equation (17.19)		Equation (17.20)		Equation (17.21)	
	8 signifi- cant figures	16 signifi- cant figures	8 signifi- cant figures	16 signifi- cant figures	8 signifi- cant figures	16 signifi- cant figures
10	.7107 9124	.7107 9126	.7504 1332	.7504 1336	.7499 6570	.7499 6576
20	.7309 1225	.7309 1229	.7501 0499	.7501 0510	.7500 1037	.7500 1050
40	.7405 8644	.7405 8652	.7500 3679	.7500 3695	.7500 1488	.7500 1511
80	.7453 3101	.7453 3118	.7500 2056	.7500 2088	.7500 1515	.7500 1563
160	.7476 8057	.7476 8088	.7500 1633	.7500 1698	.7500 1471	.7500 1569
320	.7488 4952	.7488 5014	.7500 1473	.7500 1602	.7500 1380	.7500 1570
640	.7494 3209	.7494 3338	.7500 1326	.7500 1578	.7500 1171	.7500 1570
1,280	.7497 2215	.7497 2466	.7500 1091	.7500 1572	.7500 0776	.7500 1570
2,560	.7498 6523	.7498 7020	.7500 0501	.7500 1571	.7499 9892	.7500 1570
5,120	.7499 3288	.7499 4296	.7499 9551	.7500 1570	.7499 8311	.7500 1570
10,240	.7499 5908	.7499 7933	.7499 6226	.7500 1570	.7499 3755	.7500 1570
20,480	.7499 5960	.7499 9752	.7499 5960	.7500 1570	.7499 0800	.7500 1570
40,960	.7499 2532	—	.7499 2532	—	.7498 3199	—
81,920	.7498 4868	—	.7498 4868	—	.7497 1833	—

The results are given in Table 17.1. Here n is the total number of intervals used for the distance $[0, 1]$, so that $h = 1/n$. In most cases, the calculation procedure was carried out using first 8 significant figures, and then 16 significant figures. The correct value of $y(1)$ is .75001570; the effects of both relative accuracy and round-off error are clear.

17.3 Problems

17.3.1 Discuss the results of Table 17.1 in the light of the remarks of the preceding section. (The computer used is one which rounds *down* rather than rounds *off;* thus .7777 7777 7 is recorded as .7777 7777 rather than as .7777 7778.) Suppose that only the results of column 1 were available; could we estimate a good value for $y(1)$?

17.3.2. It was remarked in section 17.2 that the cumulative error in using Equation (17.14) over an interval $b - a = nh$ should be of order h. Thus, if $y^{(1)}$ denotes the numerical approximation to $y(b)$ obtained by the use of a sub-interval of size h, and $y^{(2)}$ that obtained by the use of a sub-interval of size $(h/2)$, we can expect

$$y^{(1)} \cong y(b) + ch$$
$$y^{(2)} \cong y(b) + c(h/2).$$

The coefficient c (cf. Equation (17.15)) is assumed to be the same in these two equations, and this means that $(b-a)$ should not be too large. In any event, the two equations can now be combined to yield

$$y(b) \cong 2y^{(2)} - y^{(1)}.$$

Clearly, this *Richardson extrapolation* idea can be extended to apply to cases involving higher order errors. Do so, and consider the results of Table 17.1 in this light, commenting on the effects of round-off error.

17.3.3 Compare the expected behaviors of the exact and Euler-method solutions, and discuss error estimates, for each of

(a)
$$y' = -y, \quad y(0) = 1,$$
(b)
$$y' = y, \quad y(0) = 1,$$
(c)
$$y' = y^2, \quad y(0) = 1,$$
(d)
$$y' = \sqrt{xy}, \quad y(0) = 0.$$

17.3.4 Write out an Euler procedure for the general nth-order equation (cf Problem 17.1.1). Describe how more Taylor series terms could be added. In particular, devise a higher-order process for each of

(a)
$$y'' + p(x)y' + q(x)y = r(x);$$
(b)
$$y' = f(x, y, z), \qquad z' = g(x, y, z).$$

In general, which do you think is the better way to get more accuracy: use a higher-order Taylor expansion, or use a smaller mesh? Some computer programs adjust the step size automatically, so as to try to maintain a desired accuracy; can you devise an appropriate test that such a program could apply?

17.4 Runge-Kutta Method

This is perhaps the best known of all one-step methods; it achieves the accuracy of a high-order Taylor expansion but avoids the nuisance of evaluating such derivatives as occur in Equation (17.18). We give here the details for a third-order process, and simply quote the final results for the analogous fourth-order process.

For the equation $y' = f(x, y)$, a third-order Runge-Kutta process defines the sequence of quantities

$$k_1 = hf(x_j, y_j),$$
$$k_2 = hf(x_j + \alpha_2 h, y_j + \beta_2 k_1),$$
$$k_3 = hf(x_j + \alpha_3 h, y_j + \beta_3 k_1 + \lambda_3 k_2),$$

and then uses the formula

$$y_{j+1} = y_j + \gamma_1 k_1 + \gamma_2 k_2 + \gamma_3 k_3, \tag{17.22}$$

where the coefficients $\alpha_i, \beta_i, \gamma_i, \lambda_3$ are so chosen that the expansion for $y(x_{j+1})$ as given by Equation (17.17) agrees with that for y_{j+1} obtained from Equation (17.22), up to terms of order h^3. To obtain this latter expansion, we write

$$k_1 = hf,$$
$$k_2 = h[f + f_x \alpha_2 h + f_y \beta_2 hf + \tfrac{1}{2} f_{xx} \alpha_2^2 h^2 + f_{xy} \alpha_2 h \beta_2 hf + \tfrac{1}{2} f_{yy} \beta_2^2 h^2 f^2],$$
$$k_3 = h[f + f_x \alpha_3 h + f_y(\beta_3 hf + \lambda_3 h\{f + f_x \alpha_2 h + f_y \beta_2 hf + \cdots\})$$
$$\qquad + \tfrac{1}{2} f_{xx} \alpha_3^2 h^2 + f_{xy} \alpha_3 h(\beta_3 hf + \lambda_3 hf + \cdots)$$
$$\qquad + \tfrac{1}{2} f_{yy}(\beta_3 hf + \lambda_3 hf)^2 + \cdots] + \cdots.$$

Substitution into Equation (17.22) and comparison with Equation (17.17) then leads to

$$\gamma_1 + \gamma_2 + \gamma_3 = 1, \qquad \gamma_2\alpha_2 + \gamma_3\alpha_3 = \tfrac{1}{2}, \qquad \gamma_2\beta_2 + \gamma_3(\beta_3 + \lambda_3) = \tfrac{1}{2},$$

$$\tfrac{1}{2}\gamma_2\alpha_2^2 + \tfrac{1}{2}\gamma_3\alpha_3^2 = \tfrac{1}{6}, \qquad \gamma_2\alpha_2\beta_2 + \gamma_3(\alpha_3\beta_3 + \alpha_3\lambda_3) = \tfrac{1}{3}, \qquad (17.23)$$

$$\tfrac{1}{2}\gamma_2\beta_2^2 + \tfrac{1}{2}\gamma_3(\beta_3 + \lambda_3)^2 = \tfrac{1}{6}, \qquad \gamma_3\lambda_3\alpha_2 = \tfrac{1}{6}, \qquad \gamma_3\lambda_3\beta_2 = \tfrac{1}{6}.$$

The reader may verify that α_2 and α_3 can be given arbitrary non-zero values; the remaining equations then determine the other constants. (This degree of arbitrariness is not enough to further decrease the truncation error.) The truncation error is of order h^4; the actual expression for this error is left for an exercise.

TABLE 17.2 Computed Values of $y(1)$ for $y' = x^2 - y^2$, $y(0) = 1$

	Equation (17.24)		Equation (17.26)		Equation (17.28)	
n	8 significant figures	16 significant figures	8 significant figures	16 significant figures	8 significant figures	16 significant figures
10	.7500 1682	.7500 1684	.7516 7716	.7516 7715	.7498 5414	.7498 5415
20	.7500 1574	.7500 1577	.7502 7080	.7502 7081	.7499 7522	.7499 7526
40	.7500 1562	.7500 1571	.7500 5812	.7500 5814	.7500 0549	.7500 0559
80	.7500 1553	.7500 1570	.7500 2350	.7500 2356	.7500 1300	.7500 1317
160	.7500 1539	.7500 1570	.7500 1716	.7500 1732	.7500 1473	.7500 1507
320	.7500 1507	.7500 1570	.7500 1580	.7500 1606	.7500 1492	.7500 1554
640	.7500 1439	.7500 1570	.7500 1524	.7500 1578	.7500 1435	.7500 1566
1,280	.7500 1309	.7500 1570	.7500 1445	.7500 1572	.7500 1309	.7500 1568
2,560	.7500 1065	.7500 1570	.7500 1314	.7500 1571	.7500 1065	.7500 1569
5,120	.7500 0558	.7500 1570	.7500 1092	.7500 1570	.7500 0558	.7500 1570
10,240	.7499 9544	.7500 1570	.7500 0557	.7500 1570	.7499 9544	.7500 1570
20,480	.7499 7501	.7500 1570	.7499 9569	.7500 1570	.7499 7502	.7500 1570
40,960	.7499 3442	—	.7499 7468	—	.7499 3442	—
81,920	.7498 5331	—	.7499 3423	—	.7498 5331	—

The most commonly used Runge-Kutta process is the fourth-order one; again there is some degree of arbitrariness in the coefficients. The standard version is

$$\begin{aligned} k_1 &= hf(x_j, y_j), \\ k_2 &= hf(x_j + \tfrac{1}{2}h, y_j + \tfrac{1}{2}k_1), \\ k_3 &= hf(x_j + \tfrac{1}{2}h, y_j + \tfrac{1}{2}k_2), \qquad (17.24) \\ k_4 &= hf(x_j + h, y_j + k_3), \\ y_{j+1} &= y_j + \tfrac{1}{6}[k_1 + 2k_2 + 2k_3 + k_4], \end{aligned}$$

and the error is of order h^5. The possible advantages of making other choices for the coefficients have been much discussed in the literature.

Columns 1 and 2 of Table 17.2 give the results of using Equations (17.24) on our previous trial equation $y' = x^2 - y^2$, $y(0) = 1$; again, both 8-place and 16-place calculations were used.

17.5 Problems

17.5.1 (a) Obtain the formula for the discretization error in the third-order Runge-Kutta process.

(b) Make as convenient a choice for the coefficients satisfying Equations (17.23) as you can; what desirable features result from your choice?

17.5.2 Discuss the experimental truncation and rounding errors evident in Columns 1 and 2 of Table 17.2.

17.5.3 Use Equations (17.24) to write out (no calculations should be necessary) the fourth-order Runge-Kutta equations for an nth-order differential equation system.

17.6 Multistep Processes; Instability

In solving numerically the equation $y' = f(x, y)$, a multistep process uses one or more values of $y_{j-1}, y_{j-2}, \ldots, f(x_{j-1}, y_{j-1}), f(x_{j-2}, y_{j-2})$, ..., in addition to y_j and $f(x_j, y_j)$, to calculate y_{j+1}. Also, the process may use $f(x_{j+1}, y_{j+1})$, but in this case the values of y_{j+1} must be obtained iteratively. In general, we write

$$y_{j+1} = y_j + \alpha_1 y_{j-1} + \alpha_2 y_{j-2} + \cdots \alpha_k y_{j-k} + \beta_{-1} f(x_{j+1}, y_{j+1})$$
$$+ \beta_0 f(x_j, y_j) + \beta_1 f(x_{j-1}, y_{j-1}) + \cdots + \beta_s f(x_{j-s}, y_{j-s}), \quad (17.25)$$

where the α_i and β_i are coefficients chosen to minimize discretization error, and perhaps also to avoid computational instability.

If the term involving $f(x_{j+1}, y_{j+1})$ does appear, Equation (17.25) is an implicit equation for y_{j+1}. To solve it, we obtain a first approximation to y_{j+1} by Euler's method or by some other method, then substitute this first approximation into the right-hand side of Equation (17.25) so as to obtain an improved approximation; the process now continues iteratively. Of course, Newton's method, or some other rapidly convergent method, could be used to speed up the iteration.* Since we first predict a value for y_{j+1} and then "correct" it, computational processes in which β_{-1} is nonzero are often said to be of the "predictor-corrector" type. It is also possible to include a term in $f(x_{j+2}, y_{j+2})$ on the right-

---
*For a discussion of iterative methods, see C. Pearson, *Numerical Methods in Engineering and Science*, Van Nostrand, 1986.

hand side, but the resulting gain in accuracy may be more than offset by the increased complexity of the iterative process.

A simple multistep process for the equation $y' = f(x, y)$ is given by the *centered difference equation*

$$y_{j+1} - y_{j-1} = 2hf(x_j, y_j). \qquad (17.26)$$

The truncation error is of order h^3, rather than of order h^2 as was the case for Euler's method. Columns 3 and 4 of Table 17.2 give the results for the solution of $y' = x^2 - y^2$ by use of Equation (17.26), and the reader should again examine the effects of both truncation error and rounding error. Equation (17.26) is not "self-starting;" even if y_0 is given, it is clear that Equation (17.26) cannot be used until we have first computed y_1 by some other process. For the case of the trial problem of Table 17.2, y_1 was calculated by use of the first three terms of a Taylor expansion.

It is a typical feature of multistep methods that they are not self-starting. Another feature is the fact that a special calculation must be made if it is desired to change the step size part way through a calculation.

A multistep method may encounter a difficulty of a more serious kind which may be illustrated by examination of Equation (17.26). Suppose we consider the special case $f(x, y) = -y$, so that with $y(0) = 1$ the true solution of $y' = f$ is e^{-x}. The exact solution of the difference equation (17.26), that is, of

$$y_{j+1} - y_{j-1} = -2hy_j, \qquad (17.27)$$

has the form

$$A(-h + \sqrt{1 + h^2})^j + B(-h - \sqrt{1 + h^2})^j,$$

where the constants A and B depend on the prescribed value of y_0, and on the value of y_1 which is determined by the starting process. It is clear that as j becomes large, the second term also becomes large, and the calculation loses meaning. Even if the values of y_0 and y_1 are such that $B \equiv 0$, the effect of round-off error will be to introduce a small amount of the second of the above two components, and once introduced this component will grow without bound as j increases. Thus Equation (17.26) is basically *unstable*, at least for examples like $f = -y$.

The general case of Equation (17.26) is also unstable, in the sense that a slight change in initial conditions may have a large effect on the solution. For if we replace y_j in Equation (17.26) by $y_j + \delta y_j$, where δy_j is the resulting perturbation in y_j, then to first order we have

$$\delta y_{j+1} - \delta y_{j-1} = 2hf_y(x_j, y_j)\cdot \delta y_j.$$

If we assume that h is sufficiently small that f_y is approximately constant over a range of j values, then an argument of the kind used for Equation (17.27) demonstrates instability. Of course, the "initial conditions" could refer to any mesh point in the interval, and δy_j could result from round-off error. It is worth remarking that, in using Equation (17.26), we have approximated the first-order equation $y' = f(x, y)$ by a second-order difference equation, so that it is not surprising to find that we have introduced a spurious second solution.

At this point, the reader is probably wondering why, if Equation (17.26) is always unstable, we were able to get good results by its use for $y' = x^2 - y^2$, as demonstrated in columns 3 and 4 of Table 17.2. We hope that this apparent paradox will bother him enough that he will pause at this point and resolve it (the answer has nothing to do with the special form $f = x^2 - y^2$).

In choosing as a general criterion for instability the drastic growth with increasing j of a perturbation δy_j, we must be a little careful in our interpretation of the word "drastic." In using Euler's method on $y' = y$, $y(0) = 1$, a perturbation δy_j satisfies $\delta y_{j+1} = \delta y_j(1 + h)$ so that δy_j tends to grow as $(1+h)^j$; however, since the true solution tends to grow as e^{hj}, the *fractional* error due to the perturbation should not increase. Thus we would not describe Euler's method as (relatively) unstable for the case $f(x, y) = y$.

Returning now to Equation (17.25), let us choose as an example of a predictor-corrector multistep method for $y' = f(x, y)$ the formula

$$y_{j+1} = y_j + \tfrac{1}{2}h[f(x_j, y_j) + f(x_{j+1}, y_{j+1})]. \tag{17.28}$$

Using the formula

$$f(x_{j+1}, y_{j+1}) = f(x_j, y_j) + f_x(x_j, y_j)\cdot h + f_y(x_j, y_j)\cdot \{hf + \cdots\} + \cdots,$$

the reader may verify that the truncation error in Equation (17.28) is of order h^3. Columns 5 and 6 of Table 17.2 give the results of using Equation (17.28) for the previous problem; viz., $y' = x^2 - y^2, y(0) = 1$. Here Euler's method was used as predictor, and improved values of y_{j+1} were obtained by substituting the current approximation for y_{j+1} into the right-hand side of Equation (17.28); a total of 5 iterations was used for each value of j. Again, the reader is asked to examine the truncation error and round-off error exhibited in these two columns of Table 17.2.

A large number of multistep methods of the form (17.25) have been proposed in the literature and some general results concerning the relationship between truncation error and stability have been obtained. We will not pursue these methods further here; however, some of the following exercises provide pertinent information.

17.7 Problems

17.7.1 What is the most advantageous choice of relative truncation errors in the "predictor" and "corrector" formulas for solving Equation (17.25)? Can the predicted and corrected results be used for a criterion for the alteration of h?

17.7.2 What is the highest order of truncation error that you can achieve in Equation (17.25), if $k = s = 2$, and if the process is to be stable?

17.7.3 Devise a numerical process for solving $y' = f(x, y)$ in which *all* values of y_j are first calculated by some method, and subsequently all of these values are corrected in some manner. (Repeated correction sweeps may be used.) Discuss accuracy and stability.

17.7.4 In order to monitor accuracy for a given one-step method, it is proposed that a backwards step be taken on occasion, so as to compare new and old values of y_j. Discuss.

17.7.5 Let $y(t)$ and $z(t)$ satisfy the coupled pair of differential equations

$$y' = -133y + 66z$$
$$z' = -66y + 32z$$

for $t > 0$, with $y(0) = z(0) = 3$. Show that the exact solution involves the terms $\exp(-100t)$ and $\exp(-t)$. Suppose that a numerical solution of these equations is to be obtained by a first-order Euler method. After enough time has passed that the rapidly-decaying term ceases to play an appreciable role, one might think that the step interval h need only be small enough that the $\exp(-t)$ term be accurately modelled – say for example $h = .05$. Show however that this choice would lead to instability, and obtain a stability criterion governing the choice of h. This kind of differential equation problem, in which there are two or more widely-differing time constants, and in which stability requirements are governed by the rapidly-decaying terms even after they have ceased to appreciably affect the solution, is said to be *stiff*.

What would be the effect of evaluating the right-hand sides of the above equations at $(j+1)$, instead of at j, in the numerical method?

Can you invent a first-order stiff differential equation problem?

17.7.6 (a) What numerical procedure would you advocate for approximating the solution to an equation having a regular singular point at one end of the interval (e.g., $xy'' + y' + xy = 0$ in (0,1), with $y(0) < \infty$)?

(b) Devise a practical numerical technique for the case of a general differential equation operative over a semi-infinite interval, where one boundary condition requires "finiteness" at infinity. Construct a simple example.

17.8 Two-Point Boundary Conditions; Eigenvalue Problems

In the foregoing, we have for the most part assumed that enough boundary conditions were prescribed at the left-hand end of the interval to determine completely the numerical solution (i.e., we have dealt with the so-called initial value problem). If boundary conditions at both ends of the interval are prescribed, a modified strategy is called for.

The simplest idea is to choose, arbitrarily, enough of the unspecified conditions at the left-hand end to make the numerical procedure determinate, and to then adjust these choices until the prescribed conditions at the right-hand end are met. For example, let

$$y'' = f(x, y, y'), \qquad (17.29)$$

where f is a given function of (x, y, y'). Let the prescribed end conditions be $y(0) = 1$, $y(1) = 2$. We can choose any number β and construct a numerical approximation to a solution w of Equation (17.29) satisfying $w(0) = 1$, $w'(0) = \beta$, and we denote $w(1)$ by B. We then treat B as a function of β, and either by trial and error or by some more sophisticated interpolation process (e.g., Newton's method) we try to determine β so that $B = 2$. There is no difficulty in using a similar method for the type of boundary condition illustrated by $y(0) + 3y'(0) = 1, y(1) - y'(1) = 2$.

A rather different procedure is to write a set of difference equations approximating Equation (17.29) (making use of the given end point conditions) and to then use an iterative method to solve the resulting set of algebraic equations for the mesh point values y_j.

If the Equation (17.29) is linear, each of these methods becomes easier. For example, let

$$y'' + a_1(x)y' + a_0(x)y = g(x), \qquad (17.30)$$

where $a_1(x)$, $a_0(x)$, and $g(x)$ are given functions. Let $y(0) = 1$, $y(1) = 2$. If Y and Z are two numerical solutions of Equation (17.30) satisfying the conditions $Y(0) = 1$, $Y'(0) = \alpha$, and $Z(0) = 1$, $Z'(0) = \beta$, respectively, and if their values at $x = 1$ turn out to be $Y(1) = A$, $Z(1) = B$, then it is clear that the desired solution is given by

$$y = \frac{2 - B}{A - B} Y + \frac{A - 2}{A - B} Z. \qquad (17.31)$$

Alternatively, a difference equation approximation to Equation (17.30) would yield the set of $(n - 1)$ linear algebraic equations

$$\{1 - \frac{h}{2} a_1(x_j)\} y_{j-1} + \{a_0(x_j) \cdot h^2 - 2\} y_j + \{1 + \frac{h}{2} a_1(x_j)\} y_{j+1} = h^2 g(x_j)$$

$$(17.32)$$

for $j = 1, 2, \ldots, n - 1$, where $y_0 = 1$ and $y_n = 2$. An efficient way in which to solve these equations for the $(n - 1)$ unknowns y_j is to use Gaussian elimination; i.e., the first equation (containing only the two unknowns y_1 and y_2) is solved for y_1 in terms of y_2, and the result is substituted into the second equation, which may then be solved for y_2 in terms of y_3, and so on. The process terminates with the last equation since it contains only the two unknowns y_{n-2} and y_{n-1}, so that the substitution of the expression for y_{n-2} in terms of y_{n-1} permits us to determine y_{n-1}. Use of the formulas generated in the above "forward" sweep (i.e., giving y_{j-1} in terms of y_j) may now be used recursively to determine all of the unknown y_j in a "backwards" sweep. Once the y_j have been determined, we might want to correct them point by point so as to satisfy an equation of smaller truncation error than that associated with Equation (17.32); we remark that a smoothing process in which each corrected result is replaced by a weighted average of that corrected result and the previous uncorrected result can be useful in speeding up convergence and in avoiding instabilities.

There is a noteworthy case in which the Gaussian elimination method can be very advantageous. If the two complementary solutions of Equation (17.30) behave like a decaying and a rapidly increasing exponential, then the increasing exponential term may so thoroughly dominate the solution value at $x = 1$ that it is not feasible to try to adjust solution parameters at the left-hand end in order to achieve a desired result at the right-hand end. Gaussian elimination may then be very effective; for example, Equation (18.1) can be readily solved (using a variable mesh spacing) even for values of ϵ as small as 10^{-20}.

Each of these two methods can also be adapted to eigenvalue problems. If, for example, we have an equation of the form of Equation (6.1) in which λ must be determined so as to satisfy certain homogeneous conditions at the two ends of the interval, we can try either to determine λ iteratively until a solution satisfying the left-hand condition also satisfies the right-hand condition, or we can try to find λ so that a certain determinant constructed from a linear algebraic equation set vanishes. In the latter case, since the basic difference equations cannot effectively be applied to a solution function whose "ripple wave-length" is less than several times h, only the lower-order eigenvalues can reliably be obtained. It is also possible to determine the lower-order λ values by use of the Rayleigh quotient of Problem 6.1.6; the integrals can be approximated numerically, and the parameters of the assumed solution form can be determined so as to minimize the quotient. In fact, it is possible to devise computer programs which effectively search for the best values of the solution parameters even for cases in which the assumed solution form depends on the parameters in a nonlinear manner.

17.9 Problems

17.9.1 (a) For Equation (17.30), devise an alternative method to that leading to Equation (17.31), making use now of the idea of particular integral and complementary solutions.

(b) Write out the details of a method for solving Equation (17.30), based on forming an appropriate combination of special solutions, for the case $y'(0) = \frac{1}{2}$, $3y(1) - y'(1) = 2$.

17.9.2 Devise a feasible process for the case of a fourth-order nonlinear equation, in which two boundary conditions at each end are given. Describe any two-dimensional iteration process that you would use.

17.9.3 Let y and z be related by the two coupled equations

$$y'' = a_1(x)y + a_2(x)z + b_1(x)y' + b_2(x)z',$$
$$z'' = \alpha_1(x)y + \alpha_2(x)z + \beta_1(x)y' + \beta_2(x)z'. \tag{17.33}$$

Describe in detail how you could use a Gaussian elimination process to solve the linear equation set resulting from an approximation to Equations (17.33) via difference equations.

17.9.4 Choose a simple second-order linear equation having exponential type solutions, and demonstrate the validity of a preceding remark to the effect that the solution value at the right-hand end can be so thoroughly dominated by the increasing exponential that the first method of the last section is useless. Can you think of ways other than "linear algebraic equation solving" for handling this situation?

17.9.5 In using the Rayleigh quotient method to determine eigenvalues, it is necessary to have a formula for numerical integration. If it is desired to approximate $\int_a^b f(x)\,dx$ by $A_1 f(x_1) + A_2 f(x_2)$ where x_1 and x_2 are in the range $[a, b]$, what is the "best" choice for x_1 and x_2, and what are the corresponding values of the A_i? What about three points, or n points, instead of two (cf. the topic "Gaussian quadrature" in the literature)?

Suggested Reading

Henrici, P., *Discrete Variable Methods in Ordinary Differential Equations*, Wiley & Sons, 1962.

Schwarz, H. R., *Numerical Analysis*, Wiley & Sons, 1989, chapter 9.

Gear, W. C., *Numerical Initial Value Problems in Ordinary Differential Equations*, Prentice-Hall, 1971.

Singular Perturbation Methods | 18

18.1 The Boundary Layer Idea

There are many differential equations containing a small parameter ϵ, for whose treatment the methods of Chapters 9 and 10 are inappropriate, so that subtler exploitations of the smallness of ϵ are required. One category of such problems is readily illustrated by the following.

Let
$$\epsilon u''(x, \epsilon) - (2 - x^2)u(x, \epsilon) = -1 \quad \text{in} \quad |x| < 1 \tag{18.1}$$
with
$$u(-1, \epsilon) = u(1, \epsilon) = 0, \tag{18.2}$$
where $0 < \epsilon \ll 1$ and where u' denotes $\partial u(x, \epsilon)/\partial x$.

If, for example, ϵ were 10^{-10}, we might well imagine that the term $\epsilon u''$ could not possibly be large enough to contribute significantly to the balance demanded by Equation (18.1) and we might seek an approximate solution of Equation (18.1) as a solution of
$$(2 - x^2)u_1 = 1, \tag{18.3}$$
that is,
$$u(x, \epsilon) \simeq u_1(x) = 1/(2 - x^2). \tag{18.4}$$

This function is not consistent with the requirements of Equation

(18.2) and, in fact, we recognize immediately that, *since the order of the differential equation is lowered when $\epsilon u''$ is ignored*, the approximation of Equation (18.3) * precludes any possibility of coping with the boundary conditions.† It follows immediately from this failure either that Equations (18.1) and (18.2) admit no solution or that the term $\epsilon u''$ cannot be ignored. We reject the former possibility and we note that $\epsilon u''$ can be of the same order of magnitude as the nonhomogeneous term (unity) only if the "curvature," u'', is enormous. This implies that, if u'' is to play a significant role near any point, say $x = x_1$, the function u must be so "steep" near x_1 that its derivatives at x_1 are very much greater than u itself. As we shall see very soon, there are many problems in which this steepness is confined to a very narrow region and we can get a preview of the way in which this structure arises when we ask:

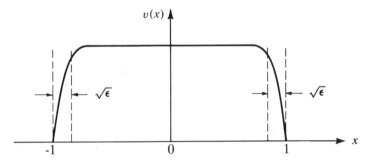

Figure 18.1 The solution $v(x)$ of the substitute problem.

what would the answer to our problem be if, instead of the positive coefficient $(2 - x^2)$, u were multiplied by the positive number a in Equation (18.1)? We think of a as any number in the range $2 > a > 1$ (since that is the range of $2-x^2$ for $|x| < 1$). The solution of this substitute problem is (see Figure 18.1)

$$v(x) = \frac{1}{a}\left(1 - \frac{\cosh x\sqrt{a/\epsilon}}{\cosh \sqrt{a/\epsilon}}\right),$$

and we see that, as we anticipated in the foregoing, $v''(x)$ is *much* greater than $v(x)$ near $x = 1$, near $x = -1$, *and nowhere else*. In fact, we see that

---

* The approximation would have been suitable, of course, in and only in the *coincidental* circumstances wherein the boundary conditions are consistent with Equation (18.3). Unfortunately, such coincidences are rare in scientific investigations.
† The reader can verify readily that this failure of Equation (18.3) to suitably approximate Equation (18.1) implies that any attempt to describe u by $u = \sum_{n=0}^{\infty} \epsilon^n u_n(x)$ will also fail.

the "boundary layers" where v is steep and $\epsilon v''$ is of order unity have a width of order $\epsilon^{1/2}$.

Led by this experience, we now ask whether the solution of Equations (18.1) and (18.2) can be approximated accurately by

$$u(x, \epsilon) = u_1(x) + p + q, \tag{18.5}$$

where p is a function which is "steep" near $x = -1$ and very small elsewhere, and q is "steep" near $x = 1$ and very small elsewhere.

To display these characteristics of p and q, we let

$$p = p(y) = p\left(\frac{x+1}{\epsilon^\alpha}\right) \quad \text{and} \quad q = q(z) = q\left(\frac{1-x}{\epsilon^\alpha}\right), \tag{18.6}$$

and we shall choose the positive number α in such a way that, for example, $p(y), p'(y), p''(y), \ldots$ are all of comparable size. Such a choice has two advantages. It implies that significant changes of p occur in distances of order unity in the y variable and, more importantly, it implies that when additive combinations of $p, p', p'', \cdots$ appear in an equation, the relative importance of such terms is displayed explicitly by the size of their coefficients.

If we substitute Equation (18.6) into (18.1) and (18.2), we obtain

$$\epsilon u_1''(x) - (2 - x^2)u_1(x) + 1 + \epsilon^{1-2\alpha}p''(y) - (2 - x^2)p(y)$$
$$+ \epsilon^{1-2\alpha}q''(z) - (2 - x^2)q(z) = 0. \tag{18.7}$$

Since we expect $p(y), q(z)$, and their derivatives to be *very* small except near $|x| = 1$, the contributions of p, p'', q, q'' cannot be significant except near $|x| = 1$. Accordingly (as we had already anticipated) $u_1(x)$ as given by Equation (18.4) is potentially an excellent approximation to a solution of Equation (18.7) away from $|x| = 1$.

Near $x = -1$, where q and q'' are very small, $2 - x^2$ can be written as

$$2 - x^2 = 1 + 2\epsilon^\alpha y - \epsilon^{2\alpha}y^2,$$

$\epsilon u_1''$ is of order ϵ compared to $p(y)$ (since $p(0) = -1$), and $\epsilon^{2\alpha}y^2 \ll 2\epsilon^\alpha y \ll 1$ in the range y of interest (since we know that $\alpha > 0$). Thus Equation (18.7) is well approximated by

$$\epsilon^{1-2\alpha}p''(y) - p(y) = 0. \tag{18.8}$$

If we choose $\alpha > \frac{1}{2}$, then Equation (18.8) degenerates to $p(y) = 0$ and we are back where we started; alternatively, if $\alpha < \frac{1}{2}$, the first term dominates Equation (18.8) and we must use

$$p''(y) \simeq 0$$

instead of Equation (18.8). Thus, with $\alpha < \frac{1}{2}$, there is no $p(y)$ which is small far from $y = -1$ except $p(y) \equiv 0$. Clearly, then, *the only acceptable choice for α is $\alpha = \frac{1}{2}$*. With that choice we find that

$$p = Ae^{-y} + Be^{y}. \tag{18.9}$$

Clearly, B must be zero in order that $p(y)$ be very small except in $0 < y < O(1)$ and, in fact, we shall always interpret the words "very small elsewhere" in the paragraph preceding Equation (18.6) to mean

$$p(y) \to 0 \quad \text{as} \quad y \to \infty.$$

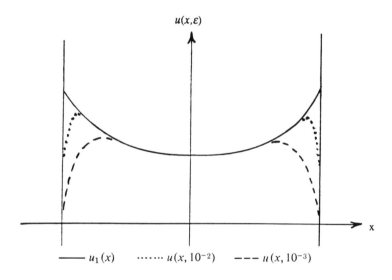

Figure 18.2

With $B = 0$, we determine A by noting that

$$u(-1) \simeq u_1(-1) + p(0) = 1 + A = 0,$$

so that

$$A = -1. \tag{18.10}$$

The symmetry of the problem is such that we see immediately that q plays the same role near $x = 1$ that p plays near $x = -1$. Thus, we have

$$u(x, \epsilon) \simeq \frac{1}{2 - x^2} - \exp\left(-\frac{x + 1}{\sqrt{\epsilon}}\right) - \exp\left(-\frac{1 - x}{\sqrt{\epsilon}}\right), \tag{18.11}$$

and this function is depicted in Figure 18.2 for each of two values of ϵ. The reader should verify carefully that, for every x in $|x| < 1$, each term of Equation (18.7) that we have ignored is small compared to those retained. This certainly doesn't *prove* anything, but one can afford to be rather confident that Equation (18.11) provides an excellent approximate description of $u(x, \epsilon)$.

We can now see that, in some problems where the differential equation $L(u, \epsilon) = 0$ contains a small parameter, ϵ, in such a way that, when $\epsilon \to 0$, the order of the differential equations is lowered:

(1) there is hope that the solution $u(x, \epsilon)$ is well approximated by $u_1(x)$, a solution of $L(u_1, 0) = 0$, except near points x_1 and x_2 at which boundary conditions are prescribed for $L(u, \epsilon) = 0$;

(2) it may be convenient to write

$$u(x, \epsilon) \simeq u_1(x) + f_1\left(\frac{x - x_1}{\epsilon^\nu}\right) + f_2\left(\frac{x - x_2}{\epsilon^\mu}\right),$$

where, because of the change of coordinate, one can readily sort out the simplest differential equation for each of the functions f_1 and f_2 which will give a suitable approximation for $u(x, \epsilon)$.

Before turning to some exercises, a mild but *nonlinear* extension of the foregoing problem may help to consolidate the ideas.
Let

$$\epsilon u''(x, \epsilon) + (1 - x^2)u(x, \epsilon) + u^2(x, \epsilon) = 1 \tag{18.12}$$

with

$$u(-1, \epsilon) = u(1, \epsilon) = 0, \qquad 0 < \epsilon \ll 1. \tag{18.13}$$

We seek a one-signed * solution of Equations (18.12) and (18.13). Except near $|x| = 1$, under the reasoning used on the foregoing problem, $u(x, \epsilon)$ *may* be well approximated by a solution of

$$u^2 + (1 - x^2)u = 1. \tag{18.14}$$

That is, $u(x, \epsilon)$ may resemble

$$u_1(x) = \tfrac{1}{2}\{x^2 - 1 + \sqrt{5 - 2x^2 + x^4}\}$$

or

$$u_2(x) = \tfrac{1}{2}\{x^2 - 1 - \sqrt{5 - 2x^2 + x^4}\}.$$

---

* That is, either $u(x) > 0$ for all x in $|x| < 1$ or $u(x) < 0$ for all x in $|x| < 1$. We specify a one-signed solution because, as we shall see later, the absence of this constraint leaves the problem with many solutions.

It is also possible that Equations (18.12) and (18.13) admit two solutions, one resembling u_1, and one resembling u_2.

Thus, we *try* (using the symmetry from the outset)

$$u(x, \epsilon) = u_1(x) + w(y) + w(z), \tag{18.15}$$

where

$$y = (x + 1)/\epsilon^\nu, \quad z = (1 - x)/\epsilon^\nu, \quad \nu > 0.$$

Near $y = 0$, we have immediately (ignoring contributions from $w(z)$)*

$$\epsilon u_1''(x) + \epsilon^{1-2\nu} w''(y) + 2\epsilon^\nu y w(y) - \epsilon^{2\nu} y^2 w(y)$$
$$+ w^2(y) + 2u_1(-1 + \epsilon^\nu y) w(y) = 0. \tag{18.16}$$

We choose $\nu = \tfrac{1}{2}$ again and we note that

$$u_1(-1 + \epsilon^\nu y) = u_1(-1) + \epsilon^{1/2} y u_1'(-1) + \cdots, \tag{18.17}$$

so that an excellent approximation to Equation (18.16) for $y = O(1)$ is

$$w'' + 2u_1(-1)w + w^2 = 0, \tag{18.18}$$

where

$$w(\infty) = 0. \tag{18.19}$$

Because of (18.13), we have

$$w(0) = -u_1(-1) = -1. \tag{18.20}$$

Note that any other choice of ν leads to a modification of Equation (18.18) which implies that $w''(y)/w(y)$ is *not* of order unity when $\epsilon \ll 1$.

When we multiply Equation (18.18) by w' and integrate, we obtain

$$w'^2 + 2w^2 + 2w^3/3 = \text{const} = k. \tag{18.21}$$

Both $w'(y)$ and $w(y)$ must tend to zero as y gets large; hence $k \equiv 0$ and

$$w'^2 + 2w^2 + 2w^3/3 = 0. \tag{18.22}$$

Since $w(0) = -1$, Equation (18.22) implies that

$$w'^2(0) = -\tfrac{4}{3},$$

which precludes a real value of $w'(0)$. Thus, we conclude that there exists no function $w(y)$ such that Equation (18.15) is a good description of $u(x, \epsilon)$.

We now turn to the second possibility, i.e. we seek $u(x, \epsilon)$ in the form

$$u(x, \epsilon) = u_2(x) + v(y) + v(z). \tag{18.23}$$

---

*The expression in the last term of Equation (18.16) means that u_1 is to be evaluated at the point $(-1+\epsilon^\nu y)$.

The manipulations which led formerly to Equation (18.18) now lead to

$$v'' - 2v + v^2 = 0, \qquad (18.24)$$

with

$$v(\infty) = 0 \quad \text{and} \quad v(0) = 1. \qquad (18.25)$$

We continue the process (as before) to find

$$(v')^2 - 2v^2 + \tfrac{2}{3}v^3 = 0,$$

and we re-arrange this to read

$$dv/(v\sqrt{3-v}) = -\sqrt{\tfrac{2}{3}}\,dy, \qquad (18.26)$$

where we choose the minus sign,* anticipating that v will decrease monotonically as y increases. An elementary integration leads to

$$v = 12e^p/(1 + e^p)^2, \qquad (18.27)$$

where

$$p = y\sqrt{2} + 2\ln(\sqrt{3} + \sqrt{2}). \qquad (18.28)$$

Thus, Equations (18.23) and (18.27) define an excellent approximation to a one-signed solution of the problem posed.

18.2 Problems

18.2.1 Use the boundary layer idea to answer the question: For what function $u(x, \epsilon)$ is

$$\epsilon u''(x, \epsilon) - u(x, \epsilon) = \cos x \quad \text{with} \quad 0 < \epsilon \ll 1 \quad \text{and with}$$

$$u(0, \epsilon) = u\left(\frac{\pi}{2}, \epsilon\right) = 1?$$

Compare the result obtained in this way with the exact solution.

18.2.2 Repeat Problem 18.2.1 with

$$\epsilon u''(x, \epsilon) + u'(x, \epsilon) + u(x, \epsilon) = 0 \quad \text{and} \quad u(0, \epsilon) = 1, \quad u'(0, \epsilon) = 1.$$

18.2.3 Can the foregoing method be used successfully on

$$\epsilon u'' + u = \cos x \quad \text{with} \quad u(0, \epsilon) = u(L, \epsilon) = 0?$$

To what feature of the problem can this be attributed?

---

*The other sign would also lead to a function $u(x, \epsilon) = U$ which obeys Equations (18.12) and (18.13), but U would not be one-signed.

18.2.4 *Begin* to find the solution of the first illustrative problem of this chapter using

(a) a power series development,

(b) an eigenfunction expansion,

(c) a numerical integration,

(d) any other method that looks even remotely promising.

Carry each of these just far enough to enable you to make a critical appraisal of the worth of the boundary layer approach.

18.2.5 Let

$$\epsilon u''(x, \epsilon) + g(x)u'(x, \epsilon) + u(x, \epsilon) = 1 \quad \text{in} \quad 0 < x < \infty$$

with

$$g(x) \geq 1 \quad \text{in} \quad 0 \leq x < \infty, \quad \text{with} \quad 0 < \epsilon \ll 1,$$

and with

$$u(0, \epsilon) = u'(0, \epsilon) = 0.$$

Find a good approximation to $u(x, \epsilon)$.

18.2.6 Let

$$\alpha u''(x, \alpha) + (2 - x^2)u(x, \alpha) = 1 \quad \text{in} \quad -1 < x < 1$$

and

$$u(-1, \alpha) = 0, \quad u(1, \alpha) = 1.$$

Find several descriptions of $u(x, \alpha)$ so that there is no real value of α for which at least one of them does not provide a compact and useful description. (For example, when $0 < -\alpha \ll 1$, a boundary-layer description is appropriate.)

(a) How many descriptions are needed to cover, conveniently, the full range of α?

(b) Plot $u'(-1, \alpha)$ versus α for all α.

(c) Plot $u(x, \alpha)$ versus x for several "interesting" values of α.

18.2.7 One encounters difficulties with the foregoing procedure in the problem:

$$[x + \epsilon u(x, \epsilon)]u'(x, \epsilon) + u(x, \epsilon) = -2 \tag{18.29}$$

with

$$u(1, \epsilon) = 1 \quad \text{and} \quad 0 < \epsilon \ll 1.$$

Find $u(0, \epsilon)$.

If, as before, one anticipates that the contribution of $\epsilon u(x, \epsilon)$ can be ignored, he obtains as an approximation

$$u(x, \epsilon) \simeq u_1(x) = -2 + (3/x). \tag{18.30}$$

Unfortunately, $u_1(x)$ becomes indefinitely large as $x \to 0$ so that, no matter how small ϵ is, $\epsilon u(x, \epsilon)$ cannot possibly be ignored near $x = 0$. Furthermore, it would be inconvenient to describe $u(x, \epsilon)$ by

$$u(x, \epsilon) = u_1(x) + w(x/\epsilon^\nu),$$

because one then would have the intricate task wherein he must find a function w which not only contains a description of the correct behavior of $u(x, \epsilon)$ near $x = 0$, but which also cancels the singular character of $u_1(x)$. Accordingly, we modify the procedure in a way which avoids this difficulty, and which also responds to the fact that if $u_1(x)$ is a good approximation for moderately small x, then, near and at the origin, $u(x, \epsilon)$ must be large compared to unity. To accomplish this, we seek $u(x, \epsilon)$ in the form

$$u(x, \epsilon) \simeq \begin{cases} u_1(x) & \text{in } A, \\ \epsilon^\alpha w(x/\epsilon^\beta) & \text{in } B, \end{cases} \tag{18.31}$$

where A and B are overlapping regions (B contains the origin but A does not), α is a negative number such that w is of order unity in B, β is a positive number which is related to the steepness of $u(x, \epsilon)$ near $x = 0$, and w is a function which has the property that $u_1(x) \cong \epsilon^\alpha w(x/\epsilon^\beta)$ in the common parts of A and B. Clearly, α, β, A, B must be determined during the analysis. We have no assurance yet that any of this can succeed and so, to test it, we substitute $\epsilon^\alpha w(z)$ for $u(x, \epsilon)$ in Equation (18.29) to obtain

$$\epsilon^{-\beta+\alpha}(\epsilon^\beta z + \epsilon^{1+\alpha} w)w' + \epsilon^\alpha w \simeq -2. \tag{18.32}$$

The problem to which the foregoing is a preamble contains the questions:

(a) Does one have any success when he tries values of α and β such that $\beta - \alpha \neq 1$? Why?

(b) What family of functions satisfies the appropriate approximation to Equation (18.32) when $\beta - \alpha = 1$?

(c) In what region D (i.e., for what values of p, q with $\epsilon^p \ll x \ll \epsilon^q$) are $u_1(x)$ and the answers to (b) both good approximations to solutions of Equation (18.29)? Note that p and q are not *uniquely* implied.

(d) Which member w of the family found under (b) is such that $\epsilon^\alpha w$ has the same behavior as $u_1(x)$ in D?

(e) What is $u(0, \epsilon)$?

(f) What is the largest amount by which, in $0 < x < 1$, the left side of Equation (18.29) fails to be equal to -2 when Equation (18.31) is used as a definition of $u(x, \epsilon)$?

(g) For what values of ϵ is the approximation any good?

(h) Can you find $u(x, \epsilon)$ exactly and compare its implications with the conclusions reached in (a) through (g)?

18.3 Interior Layers and a Spurious Construction

Despite the great power and utility of the boundary layer approximation, both it and the more formalized method of matched asymptotic expansions can lead to spurious solutions. *The authors have never seen this occur in connection with any problem which arose in a scientific context*, but the following problem identifies and illustrates the difficulty.

Let

$$\epsilon u''(x) + u^2(x) = 1 \quad \text{in} \quad |x| < 1 \tag{18.33}$$

with

$$u(-1) = u(1) = 0.$$

The boundary layer approximation to *one* solution of this problem * is given by

$$u = -1 + \frac{12 e^{p_1}}{(1 + e^{p_1})^2} + \frac{12 e^{p_2}}{(1 + e^{p_2})^2}, \tag{18.34}$$

where

---

* Note that this problem and its solution are very closely related to those of Equations (18.12) and (18.13).

$$p_1_2 = \sqrt{2/\epsilon}\,(1 \pm x) + 2\ln(\sqrt{2} + \sqrt{3}). \tag{18.35}$$

We ask whether another solution can be found which has the form

$$u = -1 + \frac{12e^{p_1}}{(1+e^{p_1})^2} + \frac{12e^{p_2}}{(1+e^{p_2})^2} + q(\zeta), \tag{18.36}$$

where $\zeta = (x - x_0)/\sqrt{\epsilon}$, x_0 is some point, $|x_0| < 1$, and

$$1 - |x_0| \gg \sqrt{\epsilon}. \tag{18.37}$$

Note that we are extending the use of the ideas which applied to functions which were steep only near a boundary to the study of functions which are steep in a narrow region which is *not* adjacent to a boundary. This does not invalidate the *ideas* which are being used; in fact, as the reader will see, we do discover by this process that there are such solutions of Equation (18.33)

The search for $q(\zeta)$ requires the same procedure which we have already followed and we find that $q(\zeta)$ obeys the equation

$$q'' - 2q + q^2 = 0 \quad \text{in} \quad |\zeta| < \infty$$

with

$$q(\infty) = q(-\infty) = 0. \tag{18.38}$$

If such a $q(\zeta)$ exists and if $q(\zeta)$ diminishes rapidly enough as $|\zeta| \to \infty$, then we would seem to have found another solution of Equation (18.33). Such a function $q(\zeta)$ is (for *any* x_0 obeying Equation (18.37))

$$q = \frac{12e^{\sqrt{2}\zeta}}{(1+e^{\sqrt{2}\zeta})^2}. \tag{18.39}$$

The function, $q(\zeta)$, as given by Equation (18.39) does decay toward zero *with exponential rapidity* as $|\zeta| \to \infty$ and u as given by Equation (18.36) *appears* to be satisfactory.

However, with

$$u = -1 + v(x), \quad w(x) = \sqrt{\epsilon}\,v'(x),$$

we can rewrite Equation (18.33) in the form

$$\sqrt{\epsilon}\,w' = 2v - v^2 \tag{18.40}$$

and

$$\sqrt{\epsilon}\,v' = w, \tag{18.41}$$

and the integral curves (see Chapter 15) of these equations have the qualitative form shown in the phase plane diagram of Figure 18.3.

Any solution of our problem must correspond to a segment of an integral curve which starts (for $x = -1$) at a point (say L) on the line $v = 1$ and proceeds clockwise along that integral curve to another (or the same) point (L') on $v = 1$.

This end point must be such that

$$x_{\text{final}} - x_{\text{initial}} = 2 = \sqrt{\epsilon} \int_L^{L'} \frac{dw}{2v - v^2}. \tag{18.42}$$

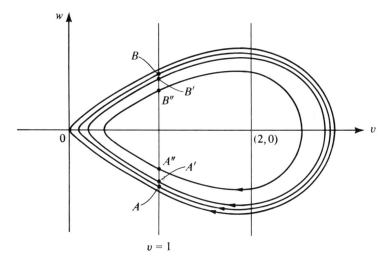

Figure 18.3 Integral curves for Equations (18.40) and (18.41).

Clearly, this is possible only for a discrete set of the integral curves. For example, the approximate solution of Equation (18.34) corresponds to the integral curve segment from A to B. A second solution corresponds, on an integral curve slightly interior to that just discussed, to a curve proceeding from B' through A' to B' and still another is the "image" of the curve from A' through B' to A'. The curve from A'' through B'' through A'' and into B'' corresponds to the solution whose approximate description is given by Equations (18.36) and (18.39), *but only if $x_0 = 0$*. There are, in fact, four solutions for which all steep regions are very close to the end points, four solutions with one steep segment in the interior (near $x = 0$), two with 2 symmetrically disposed interior steep regions, two with 2 *almost* symmetrically placed steep segments, and many others.

Thus, the solutions given by Equation (18.39) cannot be valid for most values of x_0 and we see that under very special circumstances, as in Equations (18.36) and (18.39), one can construct spurious solutions to boundary value problems using the boundary layer technique.

18.4 Problems

18.4.1 Find the boundary layer approximation to the description of the solutions identified in the discussion of Figure 18.3.

18.4.2 Imagine a three-dimensional "phase plane diagram" in connection with Equations (18.12) and (18.13) (the 3rd coordinate is x) and convince yourself that (a) this problem has many solutions, (b) plausible spurious solutions to this problem can be constructed as above.

18.4.3 Why don't spurious solutions arise in connection with the problem of Equation (18.1) or the exercises in Problems 18.2?

18.4.4 Is the *number* of solutions of Equation (18.12) and (18.13) bounded?

18.4.5 Find the periodic solutions of

$$\epsilon u''(t) - (1 - u^2)u'(t) + u(t) = 0.$$

18.4.6 Extend the ideas of Section 18.1 so that they apply to the problem

$$-\epsilon u_{yy}(x, y) + (1 + y^2)u_x(x, y) = xe^{-x} \quad \text{in} \quad |y| < 1, \quad x > 0$$

with

$$u(0, y) = u(x, -1) = u(x, 1) = 0.$$

18.5 The Turning-Point Problem

We noted in Chapter 10 that the equation

$$f''(s, \lambda) - \lambda Q(s) f(s, \lambda) = 0 \tag{18.43}$$

admits solutions whose asymptotic descriptions are given by (see Equation (10.25))

$$f_2(s, \lambda) = \exp[-\int^s h(s', \lambda) ds'], \tag{18.44}$$

where

$$h \sim \lambda^{1/2} \sum_{n=0}^{\infty} h_n(s)\lambda^{-n/2}, \tag{18.45}$$

at any s where $Q(s) \neq 0$. Conversely, at and near values of s (say s_j) where $Q(s)$ is zero, Equation (18.44) is not informative. The idea used in the earlier part of this Chapter can be exploited to find an appropriate description of u near such points, and a *uniformly valid* solution of (18.43) can be constructed. We confine our attention to functions $Q(s)$ which vanish at only one point (i.e., at one real value of s) and, without loss of generality, we choose

$$Q(s) \neq 0 \text{ when } s \neq 0; \quad Q(0) = 0; \quad Q'(0) = 1.$$

We also treat only functions Q whose power series about the origin converges in $|s| < a$.

When we use the analysis of Chapter 10 for this problem, we find that, for $h(s, \lambda)$,

$$h(s,\lambda) \sim \lambda^{1/2}\left[Q^{1/2} + \frac{1}{4\sqrt{\lambda}}\frac{Q'}{Q} - \frac{1}{\lambda}\left\{\frac{5}{32}\frac{(Q')^2}{Q^{5/2}} - \frac{1}{8}\frac{Q''}{Q^{3/2}}\right\} + \cdots\right] \tag{18.46}$$

and, for $0 < s \ll 1$,

$$h(s,\lambda) \sim \lambda^{1/2}\left[s^{1/2} + \frac{1}{4s\sqrt{\lambda}} - \frac{5}{32\lambda s^{5/2}} + \cdots\right]. \tag{18.47}$$

A comparison of the terms in this series indicates clearly that the sequence of contributions is decreasing rapidly (and is therefore a useful description) only when $|s| \gg \lambda^{-1/3}$. Thus, one can expect that Equations (18.44) and (18.46) will provide a useful description of $f(s, \lambda)$ only in $|s| \gg \lambda^{-1/3}$. Thus, if we are to find a description of f for all s, we must supplement the foregoing by an analysis which is valid (and useful) in a neighborhood of $s = 0$ which overlaps the region in which (18.46) is useful. Since this region is very narrow, it is natural to use a "stretched coordinate" for this purpose and we anticipate that the coordinate

$$\xi = s\lambda^{\nu}, \quad \nu > 0 \tag{18.48}$$

will serve our purposes. If we let

$$f(s, \lambda) = F(\xi, \lambda), \tag{18.49}$$

then, with $Q = s + q_2 s^2 + q_3 s^3 + \cdots$, Equation (18.43) becomes

$$F'' - \xi\lambda^{1-3\nu}F - q_2\xi^2\lambda^{1-4\nu}F + \cdots = 0 \tag{18.50}$$

and, when we choose $\nu = \frac{1}{3}$ and define $\epsilon = \lambda^{-1/3}$, Equation (18.50) becomes

$$F'' - \xi F - \epsilon q_2 \xi^2 F + \cdots = 0. \qquad (18.51)$$

We can also expect that, over a range of ξ such that $|\epsilon q_2 \xi^2| \ll |\xi|$, the perturbation method of Chapter 9 will provide a useful description of F. However, the foregoing inequality is equivalent to the statement

$$|\epsilon q_2 \xi| = |q_2 s| \ll 1. \qquad (18.52)$$

The region delineated by this inequality certainly overlaps that defined by $|s| \gg \lambda^{1/3}$, and the first term of the perturbation series for F should serve our purpose. Accordingly, we write

$$F = F_0(\xi) + \epsilon F_1(\xi) + \cdots$$

and obtain, from Equation (18.51),

$$F_0'' - \xi F_0 = 0. \qquad (18.53)$$

The linearly independent solutions of Equation (18.53) can be taken in the form of the Airy functions (see Chapter 11),

$$F_0 = a Ai(\xi) + b Bi(\xi), \qquad (18.54)$$

and we must find those numbers a and b for which Equations (18.54) and (18.46) have the same implications in the region of common validity [say $s = O(\lambda^{-1/6})$]. But for $s = O(\lambda^{-1/6})$, we know that $\xi \gg 1$ and (according to Chapter 11 again)

$$F_0(\xi) \simeq a \frac{1}{2\sqrt{\pi}} \xi^{-1/4} \exp\left(-\frac{2\xi^{3/2}}{3}\right)$$

plus $(b/\sqrt{\pi})\xi^{-1/4} \exp(\frac{2}{3}\xi^{3/2})$.

It follows immediately that $F_0(\xi)$ provides an approximate description of $f_2(s, \lambda)$ only if $b = 0$. Furthermore, when $s = O(\lambda^{-1/6})$, Q is well approximated by $Q \simeq s$, $h(s, \lambda)$ is given by Equation (18.47), and

$$f_2 \simeq s^{-1/4} \exp(-\tfrac{2}{3}\lambda^{1/2} s^{3/2}). \qquad (18.55)$$

Thus, $a = 2\sqrt{\pi}\lambda^{1/12}$ and $F_0(\xi)$ is given by

$$F_0(\xi) = 2\sqrt{\pi}\lambda^{1/12} Ai(\xi). \qquad (18.56)$$

For large negative values of ξ (in particular, for $-\xi = O(\lambda^{1/6})$, $F_0(\xi)$ is given by (see Chapter 11)

$$F_0(\xi) \sim \lambda^{1/12} 2(-\xi)^{-1/4} \sin\left[\frac{2}{3}(-\xi)^{3/2} + \frac{\pi}{4}\right]. \qquad (18.57)$$

The solutions of Equation (18.43) for large negative values of $\lambda^{1/3}s$ have the form

$$f(s, \lambda)$$

$$\sim (-s)^{-1/4}\left[\alpha \sin\left\{\frac{2}{3}\lambda^{1/2}(-s)^{3/2} + \frac{\pi}{4}\right\} + \beta \cos\left\{\frac{2}{3}\lambda^{1/2}(-s)^{3/2} + \frac{\pi}{4}\right\}\right],$$

(18.58)

and this agrees with Equation (18.56) only if $\beta = 0$ and $\alpha = 2$. Thus, we have

$$f_2(s, \lambda) \sim \begin{cases} Q^{-1/4}(s) \exp\left[-\lambda^{1/2}\int_0^s Q^{1/2}(s')\,ds'\right] & \text{in } s \geq 0(\lambda^{-1/6}), \\ 2\sqrt{\pi}\,\lambda^{1/12} Ai(\xi) & \text{in } |s| \leq 0(\lambda^{-1/6}), \\ 2[-Q(s)]^{-1/4} \sin\left[\lambda^{1/2}\int_0^s \sqrt{-Q(s')}\,ds' + \frac{\pi}{4}\right] \\ \qquad\qquad\qquad\qquad \text{in } -s \geq 0(\lambda^{-1/6}). \end{cases}$$

(18.59)

A second, linearly independent solution could be constructed in the same way but such a construction becomes unnecessary after the following "consolidation" of the description of the first solution, f_2.

Note that, when we define $p(s) = \int_0^s Q^{1/2}(s')\,ds'$, we obtain

$$2\sqrt{\pi}\lambda^{1/12} Ai(\lambda^{1/3}(\tfrac{3}{2}p)^{2/3}) \sim (\tfrac{3}{2}p)^{-1/6}\exp[-\lambda^{1/2}p] \quad \text{in } \lambda^{1/2}p \gg 1,$$

so that

$$2\sqrt{\pi}\,\lambda^{1/12}[\tfrac{3}{2}p(s)]^{1/6} Q^{-1/4}(s) Ai[\lambda^{1/3}(\tfrac{3}{2}p)^{2/3}] \sim Q^{-1/4}(s)\exp\left[-\lambda^{1/2}p\right].$$

Furthermore, when $s \ll 1$ (and hence $p \ll 1$), we have

$$2\sqrt{\pi}\,\lambda^{1/12}(\tfrac{3}{2}p)^{1/6} Q^{-1/4} Ai[\lambda^{1/3}(\tfrac{3}{2}p)^{2/3}] \simeq 2\sqrt{\pi}\,\lambda^{1/12} Ai(\xi)$$

and, in $-s\lambda^{-1/6} \gg 1$,

$$2\sqrt{\pi}\,\lambda^{1/12}(\tfrac{3}{2}p)^{1/6} Q^{-1/4} Ai[\lambda^{1/3}(\tfrac{3}{2}p)^{2/3}] \sim 2[-Q(s)]^{-1/4}$$
$$\times \sin\left[\lambda^{1/2}\int_0^s \sqrt{-Q(s)}\,ds'\right].$$

Therefore,

$$f_2(s, \lambda) \sim \phi_2(s, \lambda) \equiv 2\sqrt{\pi}\,\lambda^{1/12}[\tfrac{3}{2}p(s)]^{1/6} Q^{-1/4}(s) Ai\{\lambda^{1/3}[\tfrac{3}{2}p(s)]^{2/3}\}$$

(18.60)

is an asymptotic (in λ) description of f_2, valid for all real values of s.

The reader should verify that $\phi_2(s, \lambda)$ is the exact solution of

$$\phi_2'' - \lambda Q \phi_2 - \left[\frac{5}{16}\left(\frac{Q'}{Q}\right)^2 - \frac{1}{4}\frac{Q''}{Q} - \frac{5}{36}\frac{Q}{p^2}\right]\phi_2 = 0, \quad (18.61)$$

and he should note that whenever, in the context in which a given problem arises, the exact solutions of Equation (18.61) suffice * as approximations to solutions of Equation (18.43), those solutions can be written as any multiple of

$$\phi_2(s, \lambda) = p^{1/6}(s) Q^{-1/4}(s) Ai[\lambda^{1/3}(\tfrac{3}{2}p^{2/3})]$$

and

$$\phi_1(s, \lambda) = p^{1/6} Q^{-1/4} Bi[\lambda^{1/3}(\tfrac{3}{2}p)^{2/3}].$$

18.6 Problems

18.6.1 Find the most elementary, uniformly valid non-trivial descriptions of the solutions of

$$u''(x) - \lambda[\text{erf } x]u(x) = 0$$

for which, with $\lambda \gg 1$,

(a)
$$u(1) = 1, \quad u(\infty) = 0,$$

(b)
$$u(-1) = 1, \quad u(\infty) = 0,$$

(c)
$$u(-10) = u(-1) = 0,$$

(d)
$$u(-10) = u(0) = 0,$$

(e)
$$u(-10) = u(\infty) = 0.$$

18.6.2 For what large values of λ are there non-trivial solutions of

$$v''(x) + \lambda(1 - x^2)v(x) = 0$$

such that $v(\infty) = v(-\infty) = 0$?

---

* He should verify also, of course, that the coefficient of ϕ_2 in Equation (18.61) is bounded on $|x| < \infty$.

18.7 Two-timing

Each of the foregoing techniques can be regarded as a rather special case of an extremely useful expansion procedure known to some as "two-timing." The method exploits the fact that, in many problems, there appear two (or more) distinct scales in the independent variable. In the first problem of this chapter, as one can see in its answer, Equation (18.11), the two scales are unity and $\sqrt{\epsilon}$. Alternatively, in Equation (18.59), the function f_2 varies with scales which, when $s \gg \lambda^{-1/3}$, are $\lambda^{-1/2}$ and the scale of $Q(s)$; when s is very small, the scale is $\lambda^{-1/3}$. The following nonlinear problem is a useful vehicle with which to display and motivate the two-timing method.

Let

$$u''(t, \epsilon) + \epsilon(u'(t, \epsilon))^3 + u(t, \epsilon) = 0 \quad \text{in} \quad 0 < t < \infty \quad (18.62)$$

with

$$u(0, \epsilon) = 1, \quad u'(0, \epsilon) = 0 \quad (18.63)$$

and with $0 < \epsilon \ll 1$. As usual, primes denote $\partial/\partial t$. The conventional perturbation procedure (Chapter 9) is inadequate, as we shall see by using it. In that procedure we anticipate that we can describe u in the form

$$u(t, \epsilon) = u_0(t) + \epsilon u_1(t) + \cdots, \quad (18.64)$$

and the usual procedure leads to

$$u_0'' + u_0 = 0, \quad (18.65)$$

$$u_1'' + u_1 = -(u_0')^3, \quad (18.66)$$

$$\vdots$$

with $u_0(0) = 1, u_j(0) = 0, u_k'(0) = 0$, for $j > 0, k \geq 0$. These, in turn, lead to

$$u_0 = \cos t \quad (18.67)$$

and

$$u_1'' + u_1 = \tfrac{3}{4} \sin t - \tfrac{1}{4} \sin 3t, \quad (18.68)$$

so that

$$u_1 = -\tfrac{3}{8} t \cos t + \tfrac{1}{32} \sin 3t + \tfrac{9}{32} \sin t. \quad (18.69)$$

A continuation of the process clearly will lead to a series which contains a sequence of contributions of the form

$$\cos t, \quad \epsilon t \cos t, \quad \epsilon^2 t^2 \cos t, \quad \ldots,$$

and it is clear that, whether the series converges or not, it will be very difficult to interpret for $\epsilon t > 1$. Such a series is frequently said to be *nonuniformly valid* in t. In fact, there are many problems in which the series converge everywhere in t (not in this problem) and should be called *nonuniformly interpretable*, or *nonuniformly useful*.

The result of the foregoing expansion procedure suggests that the two natural scales in t are unity (because of the appearance of the function $\cos t$) and ϵ^{-1} (because of the form of the series). Accordingly, we seek a description of $u(t, \epsilon)$ in the form

$$u(t, \epsilon) = w(t, \tau, \epsilon),$$

where $\tau = \epsilon t$. Denoting a partial derivative by a subscript, the initial conditions of Equation (18.63) become

$$w(0, 0, \epsilon) = 1, \quad w_t(0, 0, \epsilon) + \epsilon w_\tau(0, 0, \epsilon) = 0, \quad (18.70)$$

and, on substitution into (18.62), our differential equation becomes

$$w_{tt} + 2\epsilon w_{t\tau} + \epsilon^2 w_{\tau\tau} + \epsilon(w_t + \epsilon w_\tau)^3 + w = 0 \quad \text{in} \quad 0 < t, \tau < \infty. \quad (18.71)$$

In fact, if we now initiate a perturbation expansion for w, that is,

$$w(t, \tau, \epsilon) = \sum_{n=0}^{\infty} w^{(n)}(t, \tau)\epsilon^n,$$

Equation (18.71) becomes

$$w_{tt}^{(0)} + w^{(0)} = 0, \quad w_{tt}^{(1)} + w^{(1)} = -[w_t^{(0)}]^3 - 2w_{t\tau}^{(0)},$$
$$w_{tt}^{(2)} + w^{(2)} = -w_{\tau\tau}^{(0)} - 2w_{t\tau}^{(1)} - 3(w_t^{(0)})^2[w_\tau^{(0)} + w_t^{(1)}], \quad (18.72)$$

and the boundary conditions become

$$w^{(0)}(0, 0) = 1, \quad w^{(j)}(0, 0) = 0 \quad (j > 0)$$

$$(18.73)$$

$$w_t^{(0)}(0, 0) = 0, \quad w_t^{(j)}(0, 0) = -w_\tau^{(j-1)}(0, 0) \quad (j > 0).$$

Clearly, the first equation of (18.72) is satisfied by

$$w^{(0)} = A_0(\tau)\cos t + B_0(\tau)\sin t,$$

and we have from (18.73) that

$$A_0(0) = 1, \quad B_0(0) = 0.$$

At this stage we have no further constraints to apply in our search for $A_0(\tau)$ and $B_0(\tau)$ because the introduction of the extra variable has provided a flexibility which we have yet to exploit. Accordingly, we go on to the second of Equations (18.72) to obtain

$$w_{tt}^{(1)} + w^{(1)} = -2(B_0'(\tau)\cos t - A_0'(\tau)\sin t) - (B_0 \cos t - A_0 \sin t)^3$$
$$= -2[B_0' + \tfrac{3}{8}B_0(B_0^2 + A_0^2)]\cos t + 2[A_0' + \tfrac{3}{8}A_0(B_0^2 + A_0^2)]\sin t$$
$$+ \frac{3A_0^2 B_0 - B_0^3}{4}\cos 3t + \frac{3B_0^2 A_0 - A_0^3}{4}\sin 3t, \qquad (18.74)$$

and we also have

$$w^{(1)}(0,0) = 0 \quad \text{and} \quad w_t^{(1)}(0,0) = -A_0'(0).$$

It is here that we must choose $A_0(\tau)$ and $B_0(\tau)$ so that no "secular" terms which grow more rapidly than $w^{(0)}$ can appear in $w^{(1)}$. To do this, we must suppress terms in $w^{(1)}$ of the form $t\cos t$ and $t\sin t$. However, such terms *must* appear unless the coefficients of $\cos t$ and $\sin t$ on the right-hand side of Equation (18.74) each vanish for all τ. Thus

$$B_0' + \tfrac{3}{8}B_0(B_0^2 + A_0^2) = 0 \quad \text{and} \quad A_0' + \tfrac{3}{8}A_0(B_0^2 + A_0^2) = 0.$$

Since we know that $A(0) = 1$ and $B(0) = 0$, the functions A_0, B_0 are

$$A_0 = (1 + 3\tau/4)^{-1/2}, \qquad B_0 = 0.$$

With these choices $w^{(1)}$ becomes

$$w^{(1)} = A_1(\tau)\cos t + B_1(\tau)\sin t + \frac{A_0^3(\tau)}{32}\sin 3t,$$

where

$$A_1(0) = 0 \quad \text{and} \quad B_1(0) = -\frac{3}{32}A_0^3(0) - A_0'(0) = \frac{9}{32}.$$

One can now continue the process, suppressing secular terms in $w^{(n)}$ by making appropriate choices for $A^{(n-1)}$ and $B^{(n-1)}$. When this is done, the uniformity of the growth rate of the $w^{(n)}$ terms allows the smallness of ϵ to guarantee that $w^{(0)}$ dominates the behavior of w. Thus, in this problem, a uniformly interpretable (and uniformly valid) description of w is given by

$$w \simeq w^{(0)}(t,\tau) = (1 + 3\tau/4)^{-1/2}\cos t = (1 + 3\epsilon t/4)^{-1/2}\cos t. \qquad (18.75)$$

If the expansion of Equation (18.64) had been continued, it would have been possible to sum analytically that part of the series in powers of ϵt which corresponds to the factor $(1 + 3\epsilon t/4)^{-1/2}$ in Equation (18.75).

Ordinarily, however, this is a clumsy procedure (especially for nonlinear problems) and, when the dependence on ϵt is of a somewhat less elementary character, such a summing may be prohibitively difficult.

A second, very informative illustration of the method is provided by the problem: find $u(x, \epsilon)$ such that

$$u''(x, \epsilon) + \beta \epsilon u'(x, \epsilon) + (k^2 + \epsilon \cos x)u(x, \epsilon) = 0 \qquad (18.76)$$

and

$$u(0, \epsilon) = 1, \quad u'(0, \epsilon) = 0, \qquad (18.77)$$

where $\beta \geq 0$ and $k > 0$. When $\beta = 0$, this is **Mathieu's equation**,[*] which has undergone extensive study but it is informative to inquire into the dependence of u on β as well as its dependence on k and ϵ.

As usual, we display by the notation $u(x, \epsilon)$ only those arguments whose explicit display serves an immediate purpose.

As in the foregoing problem, it is of value to initiate the conventional perturbation expansion of u (in powers of ϵ), not only because it identifies the differential operator with which we shall be concerned, but, more importantly, because it will provide a suggestion for the scale of the appropriate second variable.

We proceed initially with the case $\beta = 0$, both because the forthcoming equations are much less messy in this case, and because no information that we need is lost by that degenerate choice. For $\beta = 0$, with the expansion

$$u = u_0(x) + \epsilon u_1(x) + \cdots,$$

the reader should verify that we obtain

$$u_0 = \cos kx, \qquad (18.78)$$

$$u_1 = \frac{\cos (k+1)x}{2(2k+1)} - \frac{\cos (k-1)x}{2(2k-1)} + \frac{\cos kx}{4k^2 - 1}, \qquad (18.79)$$

and

$$u_2 = \frac{x \sin kx}{4k(4k^2 - 1)} + \frac{\cos (k+2)x}{16(k+1)(2k+1)} + \cdots. \qquad (18.80)$$

A superficial study of Equations (18.78) to (18.80) would indicate that no secular terms ($x \sin kx$) appear in u_0 and u_1 but that such terms do appear in u_2. This is indeed the case when $k \neq \frac{1}{2}$. However, when

---

[*] MacLachlan, N. W., *Theory and Application of Mathieu Functions*, Clarendon Press, Oxford, 1947.

$k = \frac{1}{2}$, then u_1 is given by the limit of Equation (18.79) as $k \to \frac{1}{2}$, and in that case,

$$u_1|_{k=1/2} = -\frac{x}{2} \sin \frac{x}{2} + \frac{1}{4}\left(\cos \frac{3x}{2} - \cos \frac{x}{2}\right). \tag{18.81}$$

Furthermore, if $|k - \frac{1}{2}| = 0(\epsilon)$, then u_1 has a magnitude which is $0(1/\epsilon)$, and the series $u_0 + \epsilon u_1 + \cdots$ is not easy to interpret. Accordingly, we infer that when $|k - \frac{1}{2}| \gg \epsilon$ and secular terms do not arise earlier in the series than u_2, we should use x and $\epsilon^2 x$ as the two independent variables. Conversely, where $|k - \frac{1}{2}| \leq 0(\epsilon)$, the appropriate variables are x and ϵx. We (or rather, you) will see later, in the exercises, that further variables should be introduced for k near or equal to $n/2$ with $n > 2$, but we proceed here, initially, with $|k - \frac{1}{2}| = 0(\epsilon)$. In fact, we use the notation*

$$k^2 = \tfrac{1}{4} + \sigma\epsilon,$$

so that Equation (18.76) becomes

$$u'' + \beta\epsilon u' + (\tfrac{1}{4} + \sigma\epsilon + \epsilon \cos x)u = 0. \tag{18.82}$$

We seek u in the form

$$u(x, \epsilon) = f(x, t, \epsilon) = f^{(0)}(x, t) + \epsilon f^{(1)}(x, t) + \cdots,$$

where $t = \epsilon x$.

The boundary conditions are

$$f(0, 0, \epsilon) = 1, \qquad f_x(0, 0, \epsilon) + \epsilon f_t(0, 0, \epsilon) = 0,$$

and we will use any further flexibility inherent in the two-timing process to suppress secular terms in f.

The foregoing procedures lead to

$$f^{(0)}(x, t) = A_0(t) \cos \frac{x}{2} + B_0(t) \sin \frac{x}{2} \tag{18.83}$$

with $A_0(0) = 1$, $B_0(0) = 0$. The equation for $f^{(1)}$ is

$$f^{(1)''} + \frac{1}{4}f^{(1)} = \left[A_0' + \frac{\beta}{2}A_0 - \left(\sigma - \frac{1}{2}\right)B_0\right]\sin\frac{x}{2}$$
$$+ \left[-B_0' - \frac{\beta}{2}B_0 - \left(\sigma + \frac{1}{2}\right)A_0\right]\cos\frac{x}{2}$$
$$- \frac{B_0}{2}\sin\frac{3x}{2} - \frac{A_0}{2}\cos\frac{3x}{2}.$$

---

*We take $|\sigma| < \frac{1}{2}$, and leave it to the reader to modify the discussion for the case $|\sigma| > \frac{1}{2}$.

Accordingly, we require that

$$A_0' + \frac{\beta}{2} A_0 + (\tfrac{1}{2} - \sigma)B_0 = 0 \tag{18.84}$$

and

$$B_0' + \frac{\beta}{2} B_0 + (\tfrac{1}{2} + \sigma)A_0 = 0, \tag{18.85}$$

so that, with $A_0(0) = 1$ and $B_0(0) = 0$,

$$A_0(t) = e^{-\beta t/2} \cosh\left(t\sqrt{\tfrac{1}{4} - \sigma^2}\right), \tag{18.86}$$

$$B_0(t) = -\sqrt{\frac{1+2\sigma}{1-2\sigma}} e^{-\beta t/2} \sinh\left(t\sqrt{\tfrac{1}{4} - \sigma^2}\right). \tag{18.87}$$

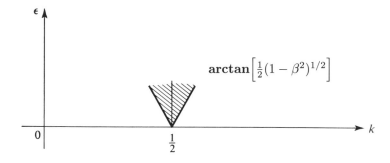

Figure 18.4 The region in k, ϵ where $u(x, \epsilon)$ is unbounded; note that the size of the region is implied by the size of β.

The procedure can be continued by finding $f_1(x, t)$, which will include terms $A_1(t)\cos(x/2)$ and $B_1(t)\sin(x/2)$; A_1 and B_1 again should be chosen to suppress all terms of the form $x\cos(x/2)$ and $x\sin(x/2)$ in $f_2(x, t)$, f_2 can then be found, and so on to exhaustion. However, as soon as one is convinced that the process *can* be continued, he concludes that $f_0(x, t)$ dominates the behavior of $u(x, \epsilon)$, and he then infers from Equations (18.83), (18.86), (18.87) that (a) $u(x, \epsilon)$ is bounded for all x if $\beta^2/4 + \sigma^2 > \tfrac{1}{4}$, (b) the solutions are unbounded in x in a region of the parameter space, k, ϵ, part of which is described in Figure 18.4. Clearly, either the "detuning" we can identify with large σ, the "damping" implied by large β, or suitable combinations of the two can prevent continued growth of $u(x, \epsilon)$.

18.8 Problems

18.8.1 (a) Without doing all the arithmetic, find a convincing demonstration that, with $\beta > 0$ and $k = 1$ or with $\beta = 0$ and $k^2 = 1 + \sigma\epsilon$ ($\sigma > 0$) or with $\beta > 0$ and $k = 1 + \sigma\epsilon$ ($\sigma > 0$), no unbounded solutions of Equation (18.76) emerge by this method.

(b) Show that, with the "damping term," $\beta\epsilon u'$, replaced by the *weaker* damping mechanism, $\beta\epsilon^2 u'$, and with $k^2 = 1 + \sigma\epsilon^2$, there are numbers $\beta > 0$, $|\sigma| > 0$ for which solutions exist that grow without bound as $x \to \infty$. Show that, for a given β, the region of the parameter space which corresponds to these growing solutions is qualitatively like that of Figure 18.5. Define that region as a function of β.

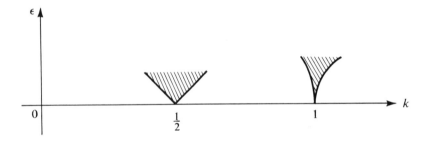

Figure 18.5 Region of parameter space corresponding to growing solutions.

(c) Show that, even with $\beta = 0$, all solutions of Equation (18.76) for which the two-timing scheme is adequate remain bounded as $x \to \infty$, when $|2k - 1| \gg \epsilon$, $|k - 1| \gg \epsilon$, and $k < 1.25$.

(d) Infer the number of appropriate independent variables for the study of solutions of Equation (18.76) for $k \simeq \frac{3}{2}$ and $\beta = 0$. For what values of the damping parameter is there a nonvanishing region of the k, ϵ space near $k = \frac{3}{2}$ in which $u(x, \epsilon)$ grows without bound as $x \to \infty$?

(e) Why did we use the words "emerge by this method" under Part (a) of this question?

(f) Characterize qualitatively and sketch the domains in the small ϵ part of the k, ϵ space in which solutions grow as $x \to \infty$.

18.8.2 When Equation (18.76) arises in a scientific context, as it frequently does, a more precise model of the phenomenon usually contains a nonlinear contribution of which an illustration is afforded by the following equation

$$u'' + \beta \epsilon u' + (k^2 + \epsilon \cos x)u + \epsilon u^3 = 0. \tag{18.88}$$

Show that the foregoing analysis leads again to a solution dominated by

$$v \simeq A_0(t) \cos \frac{x}{2} + B_0(t) \sin \frac{x}{2}$$

when $k = \frac{1}{2} + \sigma \epsilon$ and $t = \epsilon x$ but that Equation (18.84) and (18.85) are replaced by nonlinear, first-order differential equations for $A_0(t)$ and $B_0(t)$. Use the phase-plane analysis of Chapter 15 to discuss the question: Are $A_0(t)$ and $B_0(t)$ bounded? Compare the detuning effects of the ϵu^3 contribution with those of the $\epsilon \sigma u$ contribution.

18.8.3 The "method of strained coordinates" is closely related to two-timing. A classical illustration of the method concerns the non-linear oscillator problem $y'' + y = \epsilon y^3$, where $|\epsilon| \ll 1$. Let $y(0) = A$, $y'(0) = 0$. Anticipating periodic motion with period T, write $T = 2\pi(1 + \epsilon \alpha_1 + \epsilon^2 \alpha_2 + \cdots)$ where the α_i are coefficients to be determined. Choose a new independent variable τ by $\tau = \frac{2\pi}{T} t$, so that in terms of τ the period will be 2π, and now choose the α_i to eliminate secular terms. Find T. Compare this method with two-timing. What happens if the term ϵy^3 is replaced by $\epsilon |y|^3$?

A study of Problems 18.3, 18.5, and 18.8 indicates that Chapter 18 is only a cursory introduction to the techniques of singular perturbation theory. A more detailed discussion will be found in J. Kevorkian and J. D. Cole, *Perturbation Methods in Applied Mathematics*, Springer Verlag, 1981. Reference may also be made to G. F. Carrier, *Perturbation Methods*, Chapter 14 of C. Pearson (ed.), *Handbook of Applied Mathematics*, Van Nostrand, 2nd ed., 1983.

Index

admissible functions in calculus of variations, 121
adjoint operator, 68
    for difference equations, 171
Airy functions, 104, 208
asymptotic series, 86
    Airy functions, 105
    Bessel functions, 103
    gamma function, 99
    for Laplace transform, 116
    large parameter, 93
    large variable, 91
    properties, 88
    substitution into differential equation, 89

Bessel functions, 100, 103
    second kind, 101, 103
Bessel's equation, 100
Beta function, 98
boundary conditions, 8, 24
    natural, 122
    nonhomogeneous, 70, 134
    periodic, 50
boundary-layer idea, 193

centered difference equation, 186
complementary error function, 96
complementary solution, 10
continued fractions, 166
convolution integral, 114

delta function, 67
detuning, in nonlinear vibrations, 217
difference equations, 16, 157
    approximated by differential equation, 160
    constant coefficients, 158
    nonlinear, 166
    in numerical methods, 174
    second-order, 170
discretization error, 179

eigenvalue problems, 28, 47
    approximations, 75, 129
    difference equations, 172
    normalization, 50
    numerical methods, 189

perturbation methods, 82
   singular, 49
   and Sturm-Liouville theory, 53
error function, 96
Euler equation in calculus of variations, 122
Euler's constant, 108
expansion in eigenfunctions, 52, 72, 75, 132, 136, 139, 142, 172
exponential integral, 86

factored operators, 30
finite transform, 135
first-order difference equation, 16
   general solution, 18
first-order differential equation, 4, 9
Fourier series, 52, 172
fundamental solutions, 23, 24, 45
function space, 51

gamma function, 98
   asymptotic behavior, 99
   duplication formula, 98
Gaussian quadrature, 192
generating functions, for Bessel functions, 101, 165
   for difference equations, 163
   for Legendre polynomials, 106
Green's function, 64
   difference equations, 171
   symmetry, 69, 171

Hankel functions, 102, 104
heat equation, 130, 222
Helmholtz equation, 142
homogeneity, 4
   of difference equation, 16

initial condition, 8, 24
instability of numerical methods, 185
integral equation, 70

interior layers, 202
irregular singular point, 41

Laplace's equation, 136, 225
Laplace transform, 107
   asymptotic behavior, 116
   convolution formula, 114
   of derivative, 110
   of differential equation, 111
   of electric circuit, 118
   of partial differential equation, tables, 108, 111, 112, 113
Legendre polynomials, 105
Legendre's equation, 105
Liapunov function, 155
limit cycle, 155
linear dependence, 22
   and Wronskian, 43
linearity, 4
   of difference equation, 160

Mathieu's equation, 213
multiply and subtract technique, 49, 68
multistep numerical methods, 185

Newton-Raphson method, 168
node (singular point), 154
nonlinear differential equations, 13, 145
   phase plane, 148
   singular perturbations, 197
   special forms, 147
norm of function, 51
numerical methods, 174
   discretization error, 179
   eigenvalue problems, 189
   Euler method, 179, 182
   multistep methods, instability, 185
   one-step methods, 179
   quadrature, 192
   rounding errors, 181
   Runge-Kutta method, 183
   stiff problems, 189

220　Index

one-step numerical methods, 179
order of differential equation, 4
orthogonality of eigenfunctions, 49
orthonormal eigenfunctions, 50

partial differential equations, 118
    separation of variables, 130
perturbation expansion, 79
    singular, 193
phase plane, 148
Poisson equation, 137
polynomial approximation, Weierstrass' theorem, 109
power series solutions, 33

random walk problem, 159, 160
Rayleigh quotient, 76, 173, 191
Rayleigh-Ritz procedure, 78, 125
regular singular point, 41
Richardson extrapolation, 182
rounding errors of numerical methods, 181
Runge-Kutta method, 183

saddlepoint method, 99
saddle point (singular point), 155
scalar product, generalized, 51
secant method, 168
second-order differential equation, 20
    constant coefficients, 21
    nonhomogeneous, 59
self-adjoint operator, 68
separation of variables, 13
    in partial differential equations, 130

series, power, 33
    asymptotic, 86
    perturbation, 79
singular perturbation, 193
    two-timing, 210
singular points, 37, 39, 152
solution, complementary, 10
    of differential equation, 1, 10
    particular, 10
spiral point, 155
stable node, 154
stationary function in calculus of variations, 123
Stirling's formula, 99
Sturm-Liouville problem, 50, 53, 72
superposition of solutions, 10

transformation of differential equation, 20
turning-point problem, 205
two-point boundary conditions in numerical methods, 189
two-timing, 210

uniformly valid solution, 206
unstable node, 154

van der Pol equation, 152
variation of parameters, 63
variational methods, 121
    direct, 126
vortex point, 155

weighting function, 50
WKB method, 91, 93, 95, 205
Wronskian, 43
    for difference equations, 161